Food and Agriculture Organization of the United Nations

UNECE

Forest Products
Annual Market Review 2015-2016

UNITED NATIONS

NOTE

The designations employed and the presentation of material in this publication do not imply the expression of any opinion whatsoever on the part of the secretariat of the United Nations concerning the legal status of any country, territory, city or area, or of its authorities, or concerning the delimitation of its frontiers or boundaries. Data for the Commonwealth of Independent States is composed of these twelve countries: Armenia, Azerbaijan, Belarus, Georgia, Kazakhstan, Kyrgyzstan, Republic of Moldova, Russian Federation, Tajikistan, Turkmenistan, Ukraine and Uzbekistan.

DISCLAIMER

The views expressed in this paper are those of the authors and do not necessarily reflect the views or carry the endorsement of the United Nations.

ABSTRACT

The ***Forest Products Annual Market Review 2015-2016*** provides a comprehensive analysis of markets in the UNECE region and reports on the main market influences outside the UNECE region. It covers the range of products from the forest to the end-user: from roundwood and primary processed products to value-added and housing. Statistics-based chapters analyse the markets for wood raw materials, sawn softwood, sawn hardwood, wood-based panels, paper, paperboard and woodpulp. Other chapters analyse policies, pallets and wood packaging, and markets for wood energy. Underlying the analysis is a comprehensive collection of data. The Review highlights the role of sustainable forest products in international markets and discusses policies concerning forests and forest products and the main drivers and trends. The Review analyses the effects of the current economic situation on forest products markets.

ECE/TIM/SP/40

UNITED NATIONS PUBLICATIONS
Sales No. E.16.II.E.17
ISBN 978-92-1-117115-0
eISBN 978-92-1-058365-7
ISSN 1020-2269

FOREWORD

The use of forests to make wood-based products is in the midst of a paradigm shift. Environmental perspectives on harvesting trees from forests have often been moulded from bad examples, poor past practices, and deforestation occurring beyond the UNECE region. Forest harvesting and, by association, wood products have often been characterized as counter to environmental best practices, but this view is changing. In the UNECE region, forests have continued to expand and grow in volume, despite population growth and increasing demands on forests in the region. Moreover, the amenities provided by forests – such as scenery, carbon storage, clean air, wildlife habitat, watershed control and filtration, erosion protection and livelihoods (which are particularly crucial in rural areas) – are now integral parts of forest management. The majority of wood-based products sold in the UNECE region come from forests that are strictly regulated by governments to ensure their sustainability or which have independent certification confirming that the forests in which they are harvested are managed sustainably.

To say that sustainability is high on the agenda of the UN is an understatement. On 25 September 2015, world leaders adopted the 2030 Agenda for Sustainable Development, including 17 Sustainable Development Goals (SDGs), more than half of which are linked closely to forests or forest products. In addition to SDG 15 (life on land), forests and forest products play crucial roles in SDG 5 (providing affordable clean energy); SDG 6 (decent work and economic growth); SDG 7 (industry innovation and infrastructure); SDG 9 (sustainable cities and communities); SDG 10 (responsible consumption and production); and SDG 11 (action on climate change). A paradigm shift is taking place for forest products, away from the disparaging view of the recent past that the use of forests to make wood products should be minimized, to a much more positive view that sees the harvesting of forests to make wood products as a champion and key component of sustainable development.

The forest sector will play a substantial role in achieving the SDGs. For this reason, among others, the Forest Products Annual Market Review is important for the role it plays in analysing and monitoring the vitality of the sector. It is easy to see the huge potential for wood in providing sustainable energy and construction materials, but many readers of this year's special chapter on pallets and wooden packaging (such as crates and barrels) will be surprised to learn how extensively these products are used, reused and recycled in the movement of goods from producers to consumers. Many of the groceries you buy in the market arrive on wooden pallets, and, if you have a glass of fine wine, it was almost certainly aged in barrels made from sustainably harvested trees in the UNECE region.

Forest products have great promise in helping us achieve the SDGs; moreover, the global need for sustainable products and industries can transform and strengthen the forest products industries in the UNECE region. The status of forest products in the region has never been more important than it is right now. Let us seize this opportunity to foster a vibrant, sustainable sector!

Many thanks to those experts, contributors, partner organizations, information suppliers, governments and staff in our two organizations who have played a role in preparing this joint publication.

René Castro-Salazar
Assistant Director-General Forestry Department
Food and Agriculture Organization of the United Nations

Christian Friis Bach
Executive Secretary & Under-Secretary-General
United Nations Economic Commission for Europe

CONTENTS

LIST OF TABLES

LIST OF GRAPHS

ACKNOWLEDGEMENTS

The *Forest Products Annual Market Review* is the result of a cooperative effort involving a network of official country correspondents, authors, reviewers, editors, the UNECE/FAO Team of Specialists on Sustainable Forest Products and a team of people working in the Forestry and Timber Section in Geneva and in FAO, Rome. In combination, this network provides an unrivalled source of expertise and knowledge, which is the hallmark of the Review.

Many of the people involved in producing the *Forest Products Annual Market Review* volunteer their time and expertise; others are supported by companies, universities, industry associations and a variety of other organizations. Without the help of all these people and institutions, it would not be possible to produce this annual publication. This edition could not have been published without the financial support of the government of the Russian Federation.

The Review benefits from a longstanding partnership with the Department of Forest Sciences at the University of Helsinki, which provides two marketing assistants every year. This year, Teemu Anttila and Atte Koskivaara, undertook market research; prepared price data; generated all graphics and revised the Graphics Production System; and reviewed and prepared data. Their input has been critical for ensuring the quality of the publication and meeting deadlines. Professor Anne Toppinen at the Department of Forest Sciences facilitated these annual internships: we look forward to continuing this mutually beneficial arrangement.

We acknowledge the authors who wrote the chapters and, in so doing, shared their expertise and knowledge. They not only provided much of the market intelligence in their own chapters, they also assisted with data and information used elsewhere in the publication, including the summarized information in the Overview. You can find the contact details and affiliations of all authors in the annex.

The authors are as follows:

Karen Abt	Christopher Gaston	Frances Maplesden
Francisco Aguilar	Branko Glavonjić	Igor Novoselov
Eduard Akim	Antti Koskinen	Rupert Oliver
Delton Alderman	Florian Kraxner	Tapani Pahkasalo
Gunilla Beyer	Marion Le Roy	José Palacín
Ivan Eastin	Bernard Lombard	Russ Taylor
Håkan Ekström	Eugene Lopatin	Michel Valois
Kathryn Fernholz	Warren Mabee	Marshall White

In the UNECE/FAO Forestry and Timber Section, Alex McCusker collected, validated and produced the statistics; and Theresa Loeffler and Florian Steierer proofread the chapters. Karen Taylor and Anastasia Zakharova dealt with administrative matters.

The project was managed by Claudia Trentini. Matt Fonseca reviewed the chapters, provided guidance on content, and proofread the text. The chapters were reviewed at FAO by Thais Juvenal, Arvydas Lebedys, Alicja Kacprzak, Ekrem Yazici, Yanshu Li and Zuzhang Xia.

Final review in Geneva was done by Roman Michalak, Acting Chief, Forests, Land and Housing Division. Finally, the manuscript was checked by Eoin O'Driscoll, Marketing Consultant, Drima Marketing. Alastair Sarre edited the text.

In all, 42 people were directly involved in the preparation of this publication, not including the additional contributors and statistical correspondents listed separately on the following pages.

The manuscript was completed on 8 August 2016.

CONTRIBUTORS TO THE PUBLICATION

The UNECE/FAO Forestry and Timber Section would like to express its appreciation for the information and assistance received from the following people in preparing the ***Forest Products Annual Market Review 2015-2016***. Authors are acknowledged separately. The base data for the ***Review*** were supplied by country statistical correspondents, who are also acknowledged in a separate listing. We regret any omissions.

Yngve Abrahamsen, Euroconstruct, Swiss Economic Institute, Switzerland
Cristina Calderón, AEBIOM - European Biomass Association, Brussels, Belgium
Jean Christophe Claudon, International Tropical Timber Organization, Japan
Tim Cochran, Random Lengths Publication Inc, US
Ariane Crevecoeur, Confederation of European Paper Industries, Belgium
Rachel Dierolf, Sustainable Forestry Initiative, US
Gerd Ebner, Timber Online, Austria
Henric Endoff, AB Global, Sweden
Dan Meyer, Hardwood Publishing Inc, US
Alex Kuster, National Statistical Office, Switzerland
Rajmund Lackzo, Eurostat, Luxembourg
William Luppold, USDA Forest Service, US
Sundar Mahadevan, American Forest and Paper Association, US
Gabriela Sedláková, Czech Statistical Office, Czech Republic
Anne Toppinen, University of Helsinki, Finland
David Venables, American Hardwood Export Council, UK
Polly Wagner, Associated Cooperage Industries of America Inc., US
Tobias Wiegand, Studiengemeinschaft Holzleimbau, Germany
Maria Wolf-Crowther, Eurostat, Luxembourg

STATISTICAL CORRESPONDENTS

The national statistical correspondents listed below are the key suppliers of data for this publication. We are grateful for their essential contributions and their significant efforts in collecting and preparing the data.

Vardan Arevshatyan, National Statistical Service, Armenia

Guilhemine Astrie Pechin, Statistiques forestieres, Ministère de l'Agriculture, de l'alimentation, de la Pêche, de la ruralité et de l'aménagement du territoire, France

Resat Benli, General Directorate of Forestry, Turkey

Simon Bridge, Natural Resources, Canada

Kevin Borg, National Statistical Institute, Malta

Galina Ermurachi, National Bureau of Statistics, Republic of Moldova

Thomas Erhart, Office of Statistics, Liechtenstein

Spela Gale, Statistical Office, Slovenia

Branko Glavonjic, Faculty of Forestry, Belgrade State University, Serbia

Hanne Haanaes, Division for Primary Industry Statistics, Statistics Norway - Statistisk sentralbyrå, Norway

Johannes Hangler, Forest Policy and Forest Information, Federal Ministry of Agriculture, Forestry, Environment and Water Management, Austria

Antonis Horattas, Department of Forests, Ministry of Agriculture, Rural Development and Environment, Cyprus

James L. Howard, Forest Products Laboratory, USDA Forest Service, US

Surendra Joshi, Policy and Analysis Division, Swedish Forest Agency, Sweden

Boro Kovacevic, Agency for Statistics, Bosnia and Herzegovina

Nino Kverghelidze, National Statistics Office, Georgia

Elina Mäki-Simola, Statistical Services, Natural Resources Institute, Finland

Angelo Mariano, National Forest Service, Ministry of Agricultural, Food and Forest Policies, Italy

Martin Moravcik, Department of Forest Policy and Economics, National Forest Centre, Slovakia

Girgina Nikoleva, National Statistical Institute, Bulgaria

Olivian Nutescu, National Institute of Statistics, Romania

Eoin O'Driscoll, The Forest Service, Department of Agriculture, Food and the Marine, Ireland

Jan Oldenburger, Probos Foundation, Netherlands

Thierry Palgen, Administration de la nature et des forêts, Luxembourg

Ewa Ratajczak, Wood Economics Department, Wood Technology Institute, Poland

Gulaiym Shaiakhmetova, National Statistical Committee, Kyrgyzstan

Biljana Stefanova, Dissemination, State Statistical Office of the Republic of Macedonia, The former Yugoslav Republic of Macedonia

Wladyslaw Strykowski, Wood Technology Institute, Poland

Maria Torres-Quevedo Garcia de Quesada, Directorate-General of Nature Conservation, Ministry of Environment, Spain

Mati Valgepea, Department of Forestry Statistics, Estonian Environment Information Centre, Estonia

Dinko Vusic, Faculty of Forestry, University of Zagreb, Croatia

Sheila Ward, Economics and Statistics, Forestry Commission, UK

Holger Weimar, Federal Research Institute for Rural Areas, Forestry, and Fisheries, "Johann Heinrich von Thünen Institut (vTI)", Germany

Arthur Zesiger, Federal Statistical Office, Switzerland

DATA SOURCES

The data on which the ***Forest Products Annual Market Review*** is based are collected from official national correspondents through the FAO/UNECE/Eurostat/ITTO Joint Forest Sector Questionnaire, distributed in April 2016. Within the 56-country UNECE region, data for the 32 EU and EFTA countries are collected by Eurostat and for other UNECE countries by UNECE/FAO Geneva. All data are validated by UNECE/FAO Geneva.

The statistics for this ***Review*** are from the TIMBER database system. Because the database is continually updated, any single publication provides only a snapshot of the database. Data quality differs between countries, products and years. Improving data quality is a continuing task of the secretariat.

With our partner organizations and national correspondents, we strongly believe that the quality of the international statistical base for analysis of the forest products sector is improving steadily. The goal of the partner organizations is to have a complete and current database, validated by national correspondents, available from FAO in Rome, Eurostat in Luxembourg, ITTO in Yokohama and UNECE/FAO in Geneva. We are convinced that the dataset used in the ***Review*** is the best available anywhere, as of August 2016.

The data in this publication form only a small part of the total data available. ***Forest Products Statistics*** will include all available data for the years 2011-2015. The TIMBER database is available on the website of the UNECE Committee on Forests and the Forest Industry and FAO European Forestry Commission at www.unece.org/forests/fpm/onlinedata.html. More complete trade flow information is available at www.unece.org/forests/fpm/onlinedata/forest-products-trade-flow.html

The secretariat is grateful that correspondents provided actual statistics for 2015 or, in the absence of formal statistics, their best estimates. Therefore, all statistics for 2015 are provisional and subject to revision at a later date. The responsibility for national data lies with the national correspondents. The official data supplied by correspondents account for the great majority of records. In some cases, where no data were supplied or where data lacked internal consistency or were confidential, the secretariat estimated figures to keep regional and product aggregations comparable and to maintain comparability over time. Estimates are flagged in this publication, but only for products at the lowest level of aggregation.

Despite the best efforts of all involved, some significant problems remain. Chief among these are differing definitions, especially when these are not specified in the data, and unrecorded removals and production. For woodfuel removals, for example, the officially reported volumes may be as low as 20% of actual removals in some countries. The Joint Wood Energy Enquiry has gone some way towards improving the quality and coverage of data for wood energy. Conversions into the standard units used here are also not necessarily done in a consistent manner. The Joint FAO/UNECE Working Party on Forest Statistics, Economics and Management and its Team of Specialists on Forest Products Statistics is carrying out work to increase awareness of problems in measurement and how to deal with these.

In addition to the official statistics received through the Joint Forest Sector Questionnaire, trade association and government statistics have been used to complete the analysis for 2015 and early 2016. Supplementary information was obtained from experts, including national statistical correspondents, trade journals, the United Nations trade database (COMTRADE) and websites. These sources are given in the text.

EXPLANATORY NOTES

"Apparent consumption" is calculated by adding imports to a country's production and subtracting exports. Apparent consumption volumes are not adjusted for levels of stock. "Apparent consumption" is synonymous with "demand" and "use" and often referred to as "consumption". Consumption is a sum of a country's (or subregion's) production, imports and exports.

For ease of reading, the publication mostly provides value data in US dollars (indicated by the sign "$"). Unless specific for a given time period, the applied exchange rate for the euro in 2015 is €0.90 = $1 and for the Russian rouble is 60.9 RUB = $1. Both these exchange rates are based on the annual average rate provided by UNECE (http://w3.unece.org/PXWeb/en).

"Net trade" is the balance of exports and imports and is positive for net exports (i.e. when exports exceed imports) and negative for net imports (i.e. when imports exceed exports). Trade data for the 28 European Union countries include intra-EU trade, which is often estimated by the countries. Export data usually include re-exports. Subregional trade aggregates in tables include trade occurring between countries of the subregion.

For a breakdown of the region into its subregions, please see the map in the annex. References to EU28 refer collectively to the 28 country members of the EU. The term Commonwealth of Independent States (CIS) refers collectively to 12 countries: Armenia, Azerbaijan, Belarus, Georgia, Kazakhstan, Kyrgyzstan, Republic of Moldova, Russian Federation, Tajikistan, Turkmenistan, Ukraine and Uzbekistan. It is used solely for the reader's convenience.

The term "softwood" is used synonymously with "coniferous". "Hardwood" is used synonymously with "non- coniferous" or "broadleaved". More definitions appear in the electronic annex.

All references to "ton" or "tons" or «tonnes» in this text represent the metric unit of 1,000 kilograms (kg) unless otherwise indicated.

A billion refers to a thousand million (10^9).

Please note that all volumes of US and Canadian sawn softwood production and trade are given in solid m^3, converted from nominal m^3.

The use of the term "oven-dry" in this text is used in relation to the weight of a product in a completely dry state: e.g. an oven-dry metric tonne of wood fibre means 1,000 kg of wood fibre containing no moisture at all.

The term "chemical pulp" refers to semi-chemical woodpulp, chemical woodpulp and dissolving grades, unless otherwise indicated.

ACRONYMS, ABBREVIATIONS AND SYMBOLS

(Infrequently used abbreviations spelled out in the text may not be listed here)

$	US dollar
€	euro
…	not available
ACIA	Associated Cooperage Industries of America
ACP	Alternative Compliance Path
AHEC	American Hardwood Export Council
APA	The Engineered Wood Association
ARA	Amsterdam, Rotterdam and Antwerp
ATFS	American Tree Farm System
BC	British Columbia
BJC	builders' joinery and carpentry
CAD	Canadian dollar
CDM	clean development mechanism
CEPI	Confederation of European Paper Industries
CER	certified emission reduction
CETA	Comprehensive Economic and Trade Agreement
CHEP	Commonwealth Handling Equipment Pool
CHP	combined-heat-and-power
CIF	cost, insurance and freight
CIS	Commonwealth of Independent States
CLT	cross-laminated timber
CoC	chain-of-custody
COP 21	The 21st Conference of the Parties
CSA	Canadian Standards Association
DACH	the countries of Germany, Austria and Switzerland
DLH	Dalhoff Larsen and Horneman
DNA	deoxyribonucleic acid
EAB	emerald ash borer
EACU	Eurasian Customs Union
EC	European Commission
EFTA	European Free Trade Association
EPAL	European Pallet Association
EPD	environmental product declaration
EPF	European Panel Federation
EU	European Union
EUTR	European Union Timber Regulation
EWP	engineered wood products
FDA	Food and Drug Administration
FEFPEB	European Federation for Manufacturers of Wooden Packaging
FLEGT	Forest Law Enforcement, Governance and Trade
FOB	free on board
FQD	Fuel Quality Directive
FSC	Forest Stewardship Council
GDP	gross domestic product
GHG	greenhouse gas
GW	gigawatt

GWh	gigawatt hour
ha	hectare
IGI	international generic indicators
IMF	International Monetary Fund
INDC	intended nationally determined contributions
ISO	International Organization for Standardization
ITC	US International Trade Commission
ITMO	internationally transferred mitigation outcomes
ITTO	International Tropical Timber Organization
KD	kiln dried
kWh	kilowatt hour
LEED	Leadership in Energy and Environmental Design
LIRA	Leading Indicator of Remodeling Activity
LPR	La Palette Rouge
LSL	laminated strand lumber
LVL	laminated veneer lumber
m.t.	metric ton or tonne
m^2	square metre
m^3	cubic metre
MDF	medium-density fibreboard
MDG	Millennium Development Goals
MTCS	Malaysian Timber Certification System
NRV	natural range of variation
OECD	Organization for Economic Cooperation and Development
OSB	oriented strand board
OSL	oriented strand lumber
PEFC	Programme for the Endorsement of Forest Certification
PJ	petajoule
PRS	Faber Halbertsma Group Return System
PSL	parallel stand lumber
RUB	Russian rouble
SAAR	seasonally adjusted annualized rate
SAR	Special Administrative Region
SBP	Sustainable Biomass Partnership
SDG	Sustainable Development Goals
SDM	sustainable development mechanism
SFI	Sustainable Forestry Initiative
SLA	Softwood Lumber Agreement
SLB	Softwood Lumber Board
SPF	spruce, pine and fir
TPP	Trans-Pacific Partnership Agreement
TTIP	Transatlantic Trade and Investment Partnership
UK	United Kingdom of Great Britain and Northern Ireland
UNFCCC	United Nations Framework Convention on Climate Change
US	United States of America
USDA	US Department of Agriculture
USEPA	US Environmental Protection Agency
USGBC	US Green Building Council
VAT	value-added tax
VAWP	value-added wood product
VPA	voluntary partnership agreements
WTO	World Trade Organization
WWF	World Wide Fund for Nature

1 OVERVIEW OF FOREST PRODUCTS MARKETS AND POLICIES

Author of economic overview: José Palacín

HIGHLIGHTS

- The gap in the economic growth of the United States and the European Union has narrowed, as economic activity in the euro area picked up significantly in 2015. The sluggish growth trends observed in the Commonwealth of Independent States (CIS) in late 2014 worsened, however, and became an outright contraction in 2015 and early 2016.
- The general condition of forest products markets in the UNECE region was relatively stable in 2015. With the exception of the paper and paperboard industry (which contracted), consumption of all other products increased by between 1.3% and 2.6%, year-on-year.
- Active engagement by the forest sector in trade policies, environmental and climate initiatives, and voluntary programmes is contributing to increased recognition of the benefits of wood. Policy uncertainties continue, however, on forest carbon trading, sustainability rules for biomass, transatlantic trade agreements, timber-legality licensing and green building.
- About 4 billion pallets were in circulation in Europe in 2015, and about 2 billion pallets in North America. European wooden packaging producers consumed more than 20 million m^3 of sawnwood in 2015, and North American packaging producers consumed about 16.5 million m^3 of new sawnwood in 2011 and 12.6 million m^3 of reclaimed sawnwood.
- Apparent sawn softwood consumption rose by 6.1% in North America in 2015, was flat in Europe, and decreased in the CIS subregion by 2.2%.
- Sawn hardwood production in the UNECE region increased by 1.8% in 2015, to 40.7 million m^3, with production increasing in all three subregions. Apparent consumption fell in Europe and the CIS in 2015, but this was offset by rising consumption in North America.
- Wood-based panel production was stable in Europe and North America in 2015, and it increased strongly in the CIS subregion.
- Paper and paperboard production fell in Europe and North America in 2015 as graphic paper capacity continued to be closed due to the increase in reliance on electronic communication, including via the internet. Wood-pulp production rose in the CIS due to new investment into market pulp capacity.
- Expectations of increased future demand for wood pellets continue to drive growth in installed manufacturing capacity, but prices for industrial and residential pellets declined in Europe and North America in 2015 as a result of oversupply and lower demand linked to mild seasonal temperatures and low prices for competing fuel sources, especially fossil fuels.
- Housing construction in the European Union, the Russian Federation and Canada was relatively flat in 2015. The sector grew robustly in the US, although still significantly below long-term averages.

1.1 INTRODUCTION

This edition of the UNECE/FAO ***Forest Products Annual Market Review*** provides a comprehensive review of market developments in the UNECE region in 2015 through the first half of 2016 and of the policies driving those developments. The UNECE region is made up of three subregions: Europe, the Commonwealth of Independent States (CIS), and North America. It stretches from Canada and the United States of America (US) in the west through Europe to the Russian Federation and the Caucasus and Central Asian republics in the east. It includes almost all boreal and temperate forests in the Northern Hemisphere and covers about 1.7 billion hectares, which is just under half the world's total forest area.

The ***Review*** serves as a background document for the UNECE Committee on Forests and the Forest Industry, the next session of which will take place on 18-20 October 2016 in Geneva, Switzerland.

This chapter acts as an executive summary, providing an overview of the following ten chapters.

Section 1.2, which follows this section, gives a background on the macroeconomic health of the region. The effects of the economy are further elaborated on in each of the chapters, which outline the impacts of the economic situation on particular sectors and geographical regions.

The second chapter provides background on policies and market tools that are influencing the forest products sector, including those related to trade, energy and the environment (e.g. certified forest products, carbon accounting and markets, and green building).

Chapter 3 is a special chapter on pallets and wooden packaging, and chapters 4-10 cover the major forest product sectors. The Review closes with a chapter on housing, which is a leading driver of wood consumption in the UNECE region.

The Review presents and analyses the best available annual statistics for the period 2015-2016 collected by the UNECE/FAO Forestry and Timber Section from official country statistical correspondents and expert estimates.

Note that the trends discussed in this publication comprise a mix of data from the UNECE/FAO Forestry and Timber database (presented for the UNECE region as a whole and for each of the three subregions) and author-provided data, which may be derived from various sources, including the authors' own market intelligence. A strong effort has been made to reconcile data and trends, but occasionally there are differences between sources.

Additionally, there are times when authors may point to trends or data for a different geographic aggregation than the standard subregions. References to "Europe", the "CIS" and "North America" in this publication always pertain to the standard subregions.

Electronic annexes[1] provide additional statistical information and the full UNECE/FAO TIMBER database is also available on the web.[2] These comprehensive statistics, which form the basis of many of the chapters, ensure data transparency in the Review. References at the end of each chapter not only support and give credit for the ideas expressed in the chapter but also provide sources for further reading and research.

This 2016 edition of the Review also provides some analysis of markets outside the region. Forest products are increasingly traded at the global level, with pronounced effects on markets inside the UNECE region.

1.2 ECONOMIC DEVELOPMENTS WITH IMPLICATIONS FOR THE FOREST SECTOR

The world economy slowed in 2015, driven by weakened performances in emerging economies. Growth rates differed widely in the UNECE region. The gap between the pace of expansion in the US and the EU narrowed as economic activity in the euro area picked up significantly. Growth in the new EU member states was much faster than that observed in the rest of the EU, so convergence with income levels in older EU members continued, albeit at a slower rate than before the global financial crisis. In contrast, the adverse trends observed in the CIS in late 2014 became stronger, and sluggish growth was followed by an outright contraction in output. A major factor explaining the diversity of economic performances in the UNECE region was the decline in oil prices, which almost halved in 2015, with differing impacts for energy-exporting and energy-importing countries. Lower oil prices also negatively affected investment in the energy sector in some countries with advanced economies.

Job creation has been fast in the US, but wage growth has picked up only recently. As unemployment approaches levels close to what could be considered full employment, wage increases are likely to accelerate. Conditions improved in the euro area, driven by a continued output recovery in a context of moderate wage increases. The unemployment rate remains well above pre-crisis levels in the subregion, however, and the aggregate figure masks significant differences, with unemployment still very high in some countries. Labour market reforms have brought increased flexibility, but employment growth has often been dominated by low-quality jobs. The labour market in the CIS has been relatively resilient given the scale of the decline in economic activity, but unemployment has increased and remittances – which are sizeable in the poorest economies – have shrunk sharply.

Improvements in labour markets and the easing of financing conditions have boosted housing prices and residential investment in some countries with advanced economies. Construction performed particularly well in the US, soaking up pent-up demand and benefiting from acceleration in the rate of household formation. In the euro area, investment in construction turned positive after years of sustained decline. The turnaround has been driven by rising real incomes and low mortgage rates but, in some countries (such as Finland, Portugal and Spain), high levels of household debt continue to dampen construction investment. Housing prices have increased in the euro area as a whole, but they continue to fall in some countries, such as France, Finland, Greece and Italy.

---

1 www.unece.org/forests/fpamr2016-annex

2 www.unece.org/forests/fpm/onlinedata.html

Overall investment (i.e. including both residential and non-residential investment) in advanced countries has remained relatively weak amid lingering concerns over the strength of global demand, bouts of financial turbulence and, in some countries, continued high levels of indebtedness. On the other hand, easier financing conditions have facilitated deleveraging and debt refinancing, and low oil prices have been positive for corporate profitability.

The more sustained and robust recovery in the US compared with the euro area is reflected in the divergence of monetary policies. The US Federal Reserve increased rates in December 2015 after seven years of no change, but the European Central Bank announced further easing measures in March 2016. Rapid tightening in the US seems implausible given renewed concerns over growth and the easing bias signalled by monetary authorities in other advanced countries. Past expectations on the direction of monetary policy drove movements in exchange rates, with a sharp appreciation of the US dollar in 2014 dampening export performance. The euro strengthened from recent lows in relation to the US dollar in early 2016 but the appreciation has been limited. There have been large exchange-rate depreciations in the CIS as the shocks in energy-exporting countries have been transmitted to other economies in this subregion (graph 1.2.1). Some recovery in commodity prices in early 2016 was accompanied by currency strengthening.

The differing roles of fiscal policy in the US and the euro area in supporting recovery from the 2008-2009 global financial crisis help explain the differences in economic performance between the two subregions in recent years. In the euro area, past tighter fiscal policy has now been replaced by a more relaxed stance. The refugee crisis has been a source of additional fiscal outlays in a number of EU countries and is expected to provide further impetus to the construction sector. Energy exporters in the CIS were forced to undertake significant fiscal adjustments, with oil prices expected to stay low for some time.

GRAPH 1.2.1

Major currencies used to trade forest products indexed against the US dollar, January 2015-May 2016

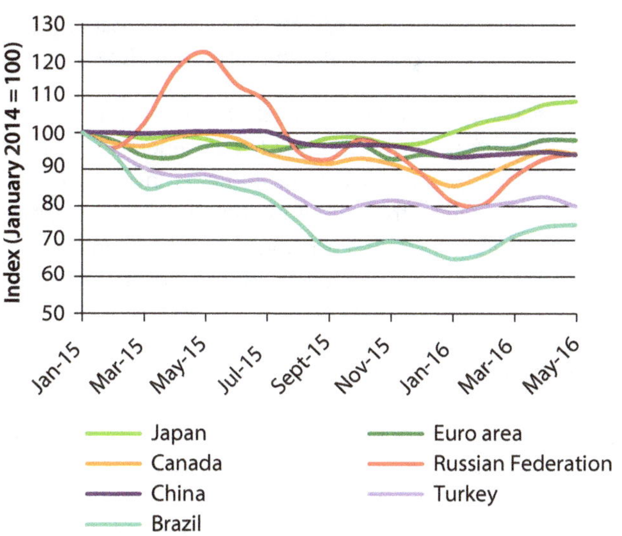

Source: IMF, 2015.

Overall, the pattern of weak recovery from the global financial crisis will continue in 2016. Some moderation is expected in the pace of economic expansion in the US and the EU, and output in the CIS is expected to shrink for a second consecutive year. Prospects are improving in line with the firming of oil prices, but vulnerability to further shocks remains.

A number of risks cloud the horizon. Geopolitical tensions, which have disrupted trade and undermined confidence in the CIS, have not yet disappeared. The refugee crisis, although providing a short-term boost to economic activity, is creating strains in a number of European countries. The UK vote to leave the EU has added another source of uncertainty that is unlikely to dissipate in the immediate future.

1.3 POLICY AND REGULATORY DEVELOPMENTS AFFECTING THE FOREST PRODUCTS SECTOR

Policy initiatives such as the EU Forest Law Enforcement, Governance and Trade (FLEGT) Action Plan, due-diligence procedures for conformance with the US Lacey Act and the EU Timber Regulation (EUTR), and the continued expansion of certified forest area worldwide demonstrate that the forest products sector is playing a leading and innovative role in the develop of a green economy.

An evaluation of the EU FLEGT Action Plan found that it had contributed to improving forest governance globally and helped reduce European imports of illegal timber. The evaluation concluded that the Plan needs to be adapted to address new challenges, particularly forest conversion, and more focus is required on private-sector engagement and communication. A review of the first two years of operation of the EUTR indicated that, although implementation has been uneven, the regulation has already contributed to significant changes in trade attitudes, structures and distribution networks.

Canada and the EU have concluded negotiations on the Comprehensive Economic and Trade Agreement (CETA), while the Transatlantic Trade and Investment Partnership (TTIP) – a trade agreement between the EU and the US – is still under negotiation. Both the CETA and the TTIP should encourage transatlantic trade, particularly in value-added forest products. The Softwood Lumber Agreement between Canada and the US has expired.

Armenia and Kyrgyzstan officially joined Belarus, Kazakhstan and the Russian Federation in the Eurasian Customs Union in 2015. The Union is designed to reduce barriers to the movement of goods, services, capital and labour among participating countries, and it is expected to increase the trade in forest products between them.

The two major certification schemes – the Forest Stewardship Council (FSC) and the Programme for the Endorsement of Forest Certification (PEFC) – report a combined global total of

462 million hectares of certified forests in 2016 (including an estimated 29.5 million hectares of overlapping dual certification), an increase of 15.8 million hectares (3.5%) over the previous 12 months. The estimated overlap of forestlands certified under both the FSC and PEFC-endorsed schemes was up by some 20 million hectares compared with previous years, when overlaps amounted to between 5 million and 8 million hectares (the large adjustment is due to improved data availability). Excluding the double-counting of forests certified under both schemes, the certified forest area worldwide amounts to 11% of the global forest area. Certified roundwood is estimated to account for about 29% of global roundwood production.

The Paris Agreement - the main outcome of the 21st Conference of the Parties to the UN Framework Convention on Climate Change (UNFCCC) - signalled that cutting emissions from deforestation and promoting sustainable forestry are now recognized globally as among the most efficient ways to address climate change.

Active engagement by the forest sector in trade policies, environmental and climate initiatives, and voluntary programmes is contributing to increased recognition of the benefits of wood. Policy uncertainties associated with ongoing debates and extended negotiations, however, have the potential to hinder further progress on several fronts, such as forest carbon trading, sustainability rules for biomass, transatlantic trade agreements, FLEGT licensing, and green building efforts. Greater alignment of these diverse efforts is needed to realize the full benefits of forests and forest products.

1.4 PALLETS AND WOODEN PACKAGING

Wooden pallets, crates and packaging cases play important roles in the movement and storage of goods worldwide. Proper design and quality standards ensure that performance is sufficient to protect the goods transported. When the economy booms, so too does the pallet and wooden packaging industry.

Pallets and wooden packaging are well positioned to flourish under policies and laws aimed at a low-carbon economy and sustainability because they have very high rates of reuse, repair and recycling, and they can be used for wood energy or the manufacture of particle board at the end of their useful lives.

Pallets are by far the most common type of wooden packaging. They provide a safe, effective transport and storage platform throughout the handling and distribution process. Pallet construction is becoming more standardized in Europe, with a migration to the EUR pallet; most prevalent is the 800 x 1,200 mm unit. Pallet sizes and designs are much less standardized in North America, where most pallets in use are custom-designed to suit transport and shipping configurations.

It is estimated that there are about 4 billion pallets in circulation in Europe and about 2 billion in North America. In Europe, the pallet and wooden packaging industry consumed more than 20

Source: E. O'Driscoll, 2016.

million m^3 of sawnwood (mostly softwood) in 2015. Similar data for North America refers to the 2011 production of just fewer than 800 million pallets (both new and rebuilt units) using about 16.5 million m^3 of new lumber and 12.6 million m^3 of reclaimed lumber. Pallet pools – whereby companies rent, lease or share the use of pallets – are increasingly used. Many companies are finding benefits in pallet "pooling", such as quality consistency, flexibility, the avoidance of capital expenditure, the reduction of costs, more cost certainty, and a reduction in the loss of assets, especially in closed pools. In North America, rental pallets are expected to facilitate the movement of food and consumer goods between the US, Canada and Mexico.

France and the US dominate global barrel exports, accounting for about $807 million and $450 million, respectively, of the value of barrel exports in 2015.

1.5 SUMMARY OF REGIONAL AND SUBREGIONAL MARKETS

The general condition of forest products markets in the UNECE region was relatively stable in 2015. With the exception of the paper and paperboard industry, consumption of all products increased by between 1.3% and 2.6%, year-on-year (table 1.5.1). There were wide differences between subregions, however.

North American and European markets experienced moderate consumption growth, benefiting from generally positive economic developments and improvements in the housing and construction industry. In contrast, deteriorating economic conditions and currency depreciations were the main reasons for a more than 4% contraction in the consumption of sawnwood and panels in the CIS countries in 2015.

As in past years, currency volatility played an important role in the trade of forest products in 2015. A strong US dollar relative to all other major currencies deeply influenced trade and ultimately production in both North America and the CIS. In particular, US imports jumped by about 10% for all wood products, while exports declined by about the same amount.

TABLE 1.5.1

Apparent consumption of industrial roundwood, sawnwood, wood-based panels and paper and paperboard in UNECE region, 2011-2015

	Thousand	2011	2012	2013	2014	2015	Change (volume) 2014-2015	Change (%) 2014-2015	Change (%) 2011-2015
Europe									
Industrial roundwood	m³	385,483	375,656	381,804	393,757	401,642	7,885	2.0	4.2
Sawnwood	m³	103,354	96,971	96,832	100,787	101,127	340	0.3	-2.2
Wood-based panels	m³	66,901	64,645	71,393	67,942	68,392	450	0.7	2.2
Paper and paperboard	m.t.	92,477	90,690	92,948	89,443	87,998	-1,445	-1.6	-4.8
CIS									
Industrial roundwood	m³	166,846	173,690	175,075	181,566	185,259	3,692	2.0	11.0
Sawnwood	m³	19,024	19,717	19,936	19,014	18,125	-889	-4.7	-4.7
Wood-based panels	m³	16,045	17,701	17,839	17,581	16,770	-810	-4.6	4.5
Paper and paperboard	m.t.	9,537	9,366	9,387	9,311	8,975	-336	-3.6	-5.9
North America									
Industrial roundwood	m³	487,212	481,158	486,764	490,150	494,286	4,136	0.8	1.5
Sawnwood	m³	89,811	95,467	101,090	106,274	112,701	6,427	6.0	25.5
Wood-based panels	m³	42,011	46,391	47,968	49,889	51,979	2,090	4.2	23.7
Paper and paperboard	m.t.	79,721	81,144	75,345	76,053	75,923	-130	-0.2	-4.8
UNECE region									
Industrial roundwood	m³	1,039,540	1,030,503	1,043,642	1,065,474	1,081,186	15,713	1.5	4.0
Sawnwood	m³	212,189	212,155	217,859	226,075	231,953	5,878	2.6	9.3
Wood-based panels	m³	124,957	128,737	137,200	135,412	137,141	1,730	1.3	9.8
Paper and paperboard	m.t.	181,734	181,200	177,680	174,807	172,896	-1,911	-1.1	-4.9

Note: Sawnwood does not include sleepers.
Source: UNECE/FAO TIMBER database, 2016.

In CIS countries, a weakened rouble pushed exports to record highs in all major product categories, in many cases more than countering the lack of domestic demand and thus increasing production.

The pulp and paper sector continues to undergo major structural changes. The increased use of electronic communication media and competitive pressure from outside the region led to reduced consumption and production in this sector in all the UNECE subregions.

1.5.1 Wood raw materials

The total consumption of roundwood – comprising logs for industrial uses and fuel – in the UNECE region was estimated at 1.28 billion m³ in 2015, an increase of 1.2% from 2014 and the third consecutive year of growth. Total log usage reached its highest level in almost ten years in each of the three UNECE subregions due to increased demand.

Removals of industrial roundwood in the UNECE region were up by 0.9% in 2015, with the biggest increase in Europe (+2%) and practically no change in North America. Although log production in the CIS subregion was only 1.1% higher in 2015 than in 2014, the longer-term trend is more impressive, with 2015 removals almost 10% above those in 2011. Almost all the increase in the timber harvest in the UNECE region in 2015 was of coniferous logs, with removals of non-coniferous logs remaining steady.

Source: **E. O'Driscoll, 2016.**

Of the total roundwood removals in the UNECE region in 2015, approximately 16% (204 million m³) were used for fuel. This volume was consumed predominantly in Europe, which accounted for almost 58% of total woodfuel consumption in the UNECE region.

The UNECE region continues to be a net exporter of logs, with globally significant trade flows of softwood logs from North America and the Russian Federation to China and the Republic of Korea. Of the top five trade flows of softwood logs worldwide, shipments to China from New Zealand, the Russian Federation and the US were all lower in 2015 compared with 2014.

Wood raw-material costs for the forest industry in the UNECE trended down for much of 2015 and in early 2016, with sawlog and pulpwood prices reaching their lowest levels (in US dollar terms) in more than six years in the first quarter of 2016. The biggest sawlog price declines in recent years have been in Austria, the Czech Republic, Germany, the Russian Federation and the western US.

Wood costs, which account for approximately 60% of the production cost in the manufacture of pulp, have fallen steadily in many UNECE countries for almost five years. Wood fibre costs have generally declined more in Europe than in North America.

1.5.2 Sawn softwood

As in 2014, 2015 saw generally mixed and unsettled global economic trends. Total sawnwood consumption in the UNECE region increased by 2.5%, from 191.5 million m³ in 2014 to 196.4 million m³ in 2015. The recovery in North America continued for the sixth consecutive year, and sawn softwood consumption increased by 6.1%. Sawnwood consumption was steady in Europe, but economic conditions and depreciating currencies in the CIS countries resulted in a decline in sawn softwood consumption of 2.2% in that subregion. The US dollar strengthened against most currencies in the first half of 2015, and volatile exchange rates affected countries differently. Sawn softwood production increased in North America by 4.2% in 2015 and in the CIS by 0.4%, while output in Europe declined by 0.7%.

The optimism and anticipated growth forecast a year ago in Europe levelled off in 2015, with apparent consumption of sawn softwood in Europe remaining at the 2014 level of 89 million m³. Economic conditions are highly variable between countries within the subregion, and this is reflected in the wide range of growth rates in softwood sawnwood consumption. Overall, Europe produced 102.6 million m³ of sawn softwood in 2015, a small decrease (0.7%) over 2014. Producers lacked an incentive to increase production, given sluggish demand in Europe and key overseas export markets.

Apparent sawn softwood consumption decreased by 2.2% in the CIS subregion in 2015, to 16.7 million m³. The production of sawn softwood was up by 0.4%, however, at an estimated 36.3 million m³. The devaluation of the rouble enabled sawn softwood producers in the Russian Federation to achieve large sales margins and high profitability, despite the strong decline in global prices (in US dollars) in the key sawn softwood export markets. Russian sawn softwood production increased slightly (by 0.2%) in 2015, to 32.1 million m³, but domestic consumption fell by 9%, to 9.8 million m³. The volume of sawn softwood exports from the Russian Federation achieved a record high of 22.4 million m³ in 2015, up by 5% from 2014, with China accounting for 44% of the Russian Federation's export volume.

Source: **UNECE/FAO, 2016.**

Apparent North American sawn softwood consumption increased by 6.1% in 2015, to 90.8 million m³. An increase in US housing starts resulted in US consumption of 75.0 million m³ in 2015 (up by 4.2%, year-on-year); in Canada, consumption reached 15.8 million m³ (up by 16.6%). US sawn softwood output was 54.3 million m³ in 2015, an increase of 1.0% compared with 2014. Canadian sawn softwood production soared by 8.3%, to 45.4 million m³. The Canadian dollar has been depreciating against the US dollar since 2014, providing Canadian producers with a competitive advantage in shipments to the US market.

1.5.3 Sawn hardwood

Sawn hardwood production in the UNECE region increased by 1.8% in 2015, to 40.7 million m³, with production increasing in all three subregions. The consumption of sawn hardwood in the UNECE region also increased, to 35.6 million m³, a 0.9% rise compared with 2014 and the fourth consecutive year of increase. Falling consumption in Europe and the CIS in 2015 was offset by rising consumption in North America.

European consumption of sawn hardwood decreased by 2.8% in 2015, to 12.2 million m³, mainly due to a significant decline in Turkish consumption. Sawn hardwood consumption in the CIS subregion fell by 25.9%, to 1.46 million m³, following a 3.6% fall in 2014. Hardwood production in the CIS increased by 2.3%, however, to 3.4 million m³, with exports taking up the slack. The weakness of the rouble encouraged a 50.5% increase in sawn hardwood exports by the Russian Federation in 2015, to 1.4 million m³. The Russian Federation exported 1.2 million m³ to China in 2015, 49% more than in 2014 and by far the highest volume ever recorded.

North American sawn hardwood consumption increased by 5.7% in 2015, to 22.0 million m³, with domestic sales in North America benefiting from rising new-home construction in the US. There are signs of slowing consumption in some sectors, however, notably pallets, flooring and board roads. US sawn hardwood exports to countries outside the subregion decreased by 8.4% in 2015, to 3.0 million m³, following five consecutive years of growth. Exports slowed to all the leading markets in 2015, including (in descending order by export volume) China (accounting for 47% of all exports), Viet Nam, Mexico, the UK, Japan and Italy.

Outside the UNECE region, China continued to dominate the sawn hardwood trade. China's imports of tropical and temperate sawn hardwoods were valued at $4.1 billion in 2015, marginally less than in 2014, when the value of imports had increased by 32% compared with 2013. Key innovations in the hardwood sector aim to extend use into new applications, notably structural applications through the development of new hardwood cross-laminated timber, glulam and laminated veneer lumber products.

1.5.4 Wood-based panels

The development of the wood-based panel sector varies significantly across the UNECE region. Plywood production in North America in 2015 (at 11.2 million m³) was almost three times higher than in both Europe and the CIS. Similarly, the production of oriented strandboard (OSB) in North America (at 18.8 million m³) was three times greater than in Europe and more than 23 times higher than production in the CIS subregion. Particle board production in Europe in 2015 (at 36.6 million m³) was four times greater than in the CIS and more than six times that in North America. Finally, medium-density fibreboard (MDF) production in Europe (at 16.3 million m³) was about four times greater than in North America and almost six times higher than in the CIS.

Overall wood-based panel production was generally stable in Europe and North America in 2015, but it was up strongly in the CIS; production trends differed substantially across product categories, however. Production was lacklustre for both particle board and plywood across the entire UNECE region. Particleboard production was down slightly in Europe and the CIS subregion and up marginally in North America. Plywood production showed little growth in Europe and the CIS and was down in North America. Production trends for OSB and MDF panels were slightly more positive, particularly in the CIS subregion, where significant increases were reported. OSB production showed moderate growth in both Europe and North America, and production was up by almost 120% in the CIS, where the OSB industry has only been established since 2012. MDF production, which was essentially stable in Europe, grew by over 3% in North America and by 26% in the CIS.

The trade situation was also interesting for wood-based panels, with currency valuations sharply affecting trade patterns across the UNECE region. For example, the weak rouble contributed to double-digit growth in exports by the CIS for all product categories, as well as to double-digit declines in panel imports in the subregion. Overall, the CIS remained a net exporter of wood-based panels in 2015. The trend was mixed in Europe, with exports of plywood and MDF up moderately and imports of plywood and OSB also rising moderately. Overall, European trade of wood-based panels was up by almost 5% compared with 2014, and Europe maintained a slight trade surplus in wood-based panels. North America had a trade deficit in wood-based panels in almost every product category; the exception was OSB, for which there was a slight trade surplus. Overall, the North American trade deficit increased by 20% in 2015.

1.5.5 Paper, paperboard and woodpulp

Paper and paperboard production fell in Europe and North America in 2015 as graphic-paper capacity continued to be reduced due to increased electronic communication, including via the internet. Wood-pulp production rose in the CIS due to new investment in market pulp capacity.

Overcapacity in the pulp, paper and paperboard segments led to closures and consolidation in 2015 and the first half of 2016; a series of strategic investments in recent years, however, mainly in tissue, specialty packaging and dissolving pulp, have breathed life into the sector. Woodpulp production in the UNECE region declined in 2015 due to closures of integrated paper machines, longer maintenance downtimes as a result of unplanned mechanical issues, and the permanent removal of market pulp capacity. Currency exchange rates and increased competition from market pulp mills outside the UNECE region were also part of the cause.

Capacity closures in graphic grades in Europe and North America continued in 2015 as a result of declining demand. Prices that had been on the decline for years finally appeared to bottom out, and a recovery was underway in early 2016. Large-scale rationalization and concentration in the paperboard sector was largely over by the end of 2015; expansions achieved by converting idled newsprint capacity were still viewed as a financially viable option, however.

Emerging markets and improved standards of living provided impetus for further expansions in sanitary and household tissue capacity in North America in 2015, while research and marketing opportunities allowed strategic investments into specialty packaging assets.

Outside the UNECE region, ongoing major investments in chemical pulp, tissue and specialty packaging papers continue to have a profound influence in Europe, North America and the CIS. Investments in research and innovation are enabling industries to thrive, but companies in the region face mounting pressures from outside competitors, many of which enjoy favourable exchange rates, advantageous production and cost structures, and financial assistance from state-run development banks and financial institutions. Such pressures on UNECE mills have led to trade sanctions, including the imposition of import tariffs.

The global pulp, paper and paperboard industry is evolving, and the challenges facing the sector in the UNECE region are vast and complex. Nonetheless, innovation, research into new products, and entrepreneurship are leading the way in maintaining an industry that is vital for many stakeholders.

Source: UNECE, 2016.

1.5.6 Wood energy

Wood energy consumption in the UNECE region shows year-on-year variations associated with fluctuations in weather patterns and competing energy sources. Expectations for future demand for wood pellets, for example, continues to drive growth in installed manufacturing capacity, but prices for industrial and residential pellets declined in Europe and North America in 2015 as a result of oversupply and lower demand linked to mild seasonal temperatures. Wood pellet trade continues to expand; North America exported 6.2 million tonnes to the EU28 in 2015, with the UK the largest importer of industrial pellets. Demand for wood energy feedstock (including pellets, briquettes and chips) is increasing in the CIS, and depreciation of the rouble might enhance opportunities for increased exports – although geopolitical dynamics might counter that effect.

The UNFCCC's Paris Agreement to curtail greenhouse gas emissions could support the expansion of wood energy use, but the role that wood will play toward meeting targets will depend on many external factors, including the carbon-accounting framework adopted by each nation. Sustainability guidelines for the use of wood for energy continue to evolve after many years of collaboration between public and private stakeholders, and standards might be adopted in the near future. Changes to the policy environment – ranging from an international climate agreement to regional renewable energy efforts such as a post-EU2020 agenda – add significant uncertainty to wood energy markets. The public policy renewable energy framework is in flux, with potential amendments to eligibility rules and financial incentive programmes. Small-scale, highly efficient power-and-heat-generation units using woodfuel present opportunities for the expansion of wood energy use; some such units seem close to commercial feasibility, but low-priced fossil fuels and the enhanced cost-effectiveness of other renewables pose significant barriers.

1.5.7 Value-added wood products

Value-added wood product markets are recovering unevenly, depending largely on housing and renovation expenditure. US import markets have grown rapidly to reach pre-global financial crisis levels, but the growth of European markets has been more moderate or even stagnant. Trade flows have been strongly affected by the strength of the US dollar against all major currencies.

Global furniture production was valued at $410 billion in 2015, down slightly from 2014. The volume of furniture trade also declined slightly in 2015, to an estimated $140 billion, but the overall trend is for increasing furniture exports. About 65% of furniture is consumed in the country in which it is manufactured and the other 35% is exported. The furniture industry has changed dramatically in recent decades, with domestic producers seeking to retain competitiveness and market share by improving their ability to serve the marketplace in terms of quality, customization and quick delivery.

US imports of builders' joinery and carpentry continued a strong increasing trend, reaching $2 billion in 2015, driven by increased spending on remodelling. European imports were essentially flat, however, with the strong devaluation of the euro translating into much lower import values in dollar terms.

Profiled-wood markets have been growing slowly. Many large suppliers to US markets have increased production capacity, while a large share of the existing production is now targeted at other markets, principally China. Profiled-wood markets in Europe declined in 2015 and are supplied mainly by regional producers.

North American production of glulam, wooden I-beams and laminated veneer lumber has made significant gains from 2010 through to the forecast for 2016, tied to increased new housing construction. European glulam production has been more uneven, with the output of some countries (e.g. Austria, Finland and Romania) heavily affected by their exports to Japan and other countries (Germany and Sweden) benefiting from sustained domestic demand.

Cross-laminated timber (CLT) production is expanding globally, with new production facilities outside the DACH countries – Germany, Austria and Switzerland – where production originally began. Global CLT production in 2015 was estimated at 650,000-700,000 m^3. Production growth in DACH countries slowed in the past year, partly because of slowing exports and softer domestic demand. In Europe, CLT construction is 10-15% more expensive than masonry and cement construction, but it is hoped that costs will come down with the further development of the industry and the standardization of products. In contrast, the use of CLT for building is quickly gaining momentum in North America, thanks in large part to interest shown in west-coast cities with strong wood cultures (e.g. Portland, Seattle and Vancouver), newly established quality and performance standards, and investments in manufacturing the product.

1.5.8 Housing

The European and North American housing markets are still recovering from the global financial crisis. In general, subdued economic forecasts for many countries with advanced economies in the UNECE region suggest that housing construction and sales in 2016 and 2017 might be similar to those reported in the past few years.

Source: proHolz, 2016.

In Europe, construction increased by 0.1% between March 2015 and March 2016, due primarily to improvements in the building construction sector, while the civil engineering sector declined, year-on-year. On a monetary basis, remodelling constitutes the largest component (about 60%) of euro-area residential construction. New residential construction is forecast to increase at a higher rate than remodelling in the immediate future. The key factors positively affecting the residential sector are financing and improved economic prospects, consistent with a slowly recovering European economy. Belgium, France, Germany, the Netherlands and Spain currently account for 73% of total residential output in the subregion, and these countries are forecast to lead housing starts in the near term. The renovation of buildings to reflect changing demand requirements and preferences in light of demographic ageing will also become an essential aspect of future housing construction.

The US housing market has continued its ascent from the 2009 trough and the Canadian market has remained steady, despite recent declines in oil prices and revenues. All sectors of the US housing market improved in 2015. Beginner or starter housing was weak, however, and the number of dwellings being built is insufficient to match population growth. Residential investment is a crucial contributor to US gross domestic product (GDP): it accounted for 6.1% of GDP in 2005 but was only 3.0% in 2014 and 3.2% in 2015, an indication that the new-home construction market has further room to expand.

Just over 280,000 residential buildings were built in the Russian Federation in 2015, a decrease of 1.1% over 2014. Overall, 418.2 million m^2 of floor space was put in place in the country in 2015, an increase of 3.4%, year-on-year; combined with a decrease in the construction of residential buildings, this indicates an increase in the average floor area of new buildings. Subdued near-term economic growth is expected to result in sizeable budget cuts and delays in residential construction programmes.

2 POLICIES SHAPING FOREST PRODUCTS MARKETS

Lead author: ***Kathryn Fernholz***

Contributing authors:

Florian Kraxner,
Igor Novoselov,
and Rupert Oliver

HIGHLIGHTS

- The Transatlantic Trade and Investment Partnership agreement between the EU and the US is still under negotiation, with an additional set of documents released in March 2016.
- Canada and the EU concluded negotiations on the Comprehensive Economic and Trade Agreement (CETA) in September 2014. Next steps include final approval and actions to ensure that policies, regulations and laws comply with CETA obligations.
- The Softwood Lumber Agreement between Canada and the US expired on 12 October 2015.
- Lumber Liquidators reached a settlement with the US Department of Justice in 2015 to plead guilty to Lacey Act violations and pay $10 million in fines, including contributions to conservation organizations.
- The two major certification schemes reported a combined global total of 462 million hectares of certified forests as of May 2016 (includes an estimated 29.5 million hectares of overlapping dual certification), an increase of 15.8 million hectares (3.5%) over the previous 12 months. Excluding the double-counting of forests certified under both schemes, the certified forest area worldwide amounts to 11% of the global forest area.
- Certified roundwood production is estimated to account for about 29% of global industrial roundwood production.
- The 21st Conference of the Parties to the UN Framework Convention on Climate Change in Paris in December 2015 raised the political profile of forests and signalled that cutting emissions from deforestation and promoting sustainable forestry are now recognized globally as among the most efficient ways to address climate change.
- The EU Emissions Trading System is the dominant world carbon market, but it continues to be affected by oversupply and low carbon prices. There were 1.7 billion surplus allowances in the market in 2015, down from 2.1 billion in 2014. The EU carbon price continues to be in the range of €5-6 per tonne.
- An evaluation of the EU FLEGT Action Plan found that it had contributed to improving forest governance globally and helped reduce European imports of illegal timber. The evaluation concluded that the Plan needs to be adapted to address new challenges, particularly forest conversion, and it should focus more on private-sector engagement and communication.
- A review of the first two years of operation of the EU Timber Regulation indicated that while implementation has been uneven, the regulation is already contributing to significant changes in trade attitudes, structures and distribution networks.
- In April 2016, the US Green Building Council announced a new credit in the LEED programme designed to address illegal wood and promote the use of verified-legal wood in building.

2.1 INTRODUCTION

Policy initiatives such as the EU Forest Law Enforcement, Governance and Trade (FLEGT) Action Plan, due diligence procedures for conformance with the US Lacey Act and the EU Timber Regulation (EUTR), and the continued expansion of certified forest area worldwide, demonstrate that the forest products sector is playing a leading and innovative role in developing a green economy. Active engagement by the sector in trade policies, environmental and climate initiatives, and voluntary programmes is contributing to increased recognition of the benefits of wood.

The Paris Agreement - the main outcome of the 21st Conference of the Parties to the UN Framework Convention on Climate Change (UNFCCC) - signalled that cutting emissions from deforestation and promoting sustainable forestry are now recognized globally as among the most efficient ways to address climate change. However, policy uncertainties associated with ongoing debates and extended negotiations have the potential to hinder further progress on several fronts, such as forest carbon trading, sustainability rules for biomass, transatlantic trade agreements, FLEGT licensing, and green building efforts.

2.2 TRADE-RELATED

2.2.1 Transatlantic free trade

Canada and the EU concluded negotiations on the Comprehensive Economic and Trade Agreement (CETA) in September 2014. On 29 February 2016, Canada's Minister of International Trade and the EU's Commissioner for Trade announced the completion of the legal review of CETA. The next steps in the process are final approval of the agreement in Canada and the EU and the actions needed to ensure that policies, regulations and laws comply with CETA obligations (GAC, 2016a).

The Transatlantic Trade and Investment Partnership (TTIP) – a trade agreement between the EU and the US – is still under negotiation, with an additional set of documents released in March 2016 (European Commission, 2016a).

Both the CETA and TTIP should encourage transatlantic trade particularly in value added forest products like prefabricated buildings, wood veneers and plywood which are subject to tariffs of up to 10%, unlike log and sawn wood products, many of which are already tariff free (GAC, 2016b). In addition to tariffs, the agreements include provisions to remove barriers to market access in public procurement procedures that go beyond existing World Trade Organization Government Procurement Agreement rules with potential implications for government timber purchasing policies on both sides of the Atlantic. There are also provisions for greater co-operation on phytosanitary issues, which are becoming increasingly important in forest products trade.

2.2.2 US and Canada forest trade and policy

The Softwood Lumber Agreement (SLA) between Canada and the US expired on 12 October 2015. In place since 2006, this agreement addresses tariffs on lumber traded between the two countries. Circumstances have changed in the last decade, including declines in US housing starts and Canadian timber supply, fluctuating lumber prices and currency exchange rates, and growth in the softwood lumber trade between Canada and China (Christensen, 2015). These and other considerations influence both the urgency and content of negotiations for a new SLA.

Pertinent to Canada's forest policy is the Canadian Boreal Forest Agreement, which addresses the management of more than 73 million hectares of the country's boreal forests (Canadian Boreal Forest Agreement, 2015). Following a report on ecosystem-based management released in August 2014 (Van Damme et al., 2014), a "natural range of variation" (NRV) strategy was released in December 2015 (Science Committee of the Canadian Boreal Forest Agreement, 2015). The concept of NRV is central to ecosystem-based management; in February 2016, member companies of the Forest Products Association of Canada announced their commitment to the "Forestry Requirements for Natural Range of Variation (NRV) Analysis and Target Setting" (Roddy, 2016), which sets the NRV of forest structure and composition as the principal guide for a variety of types and scales of management activities.

Source: UNECE. 2014.

2.2.3 CIS forest trade and policy

The Russian forest industry continues to develop in challenging economic conditions. The government has approved a set of measures to promote the forest industry (Government of the Russian Federation, 2015a), notably an anti-crisis plan of action (Government of the Russian Federation, 2015b), a new edition of the state programme, "The Development of Industry and Increase its Competitiveness" (Government of the Russian Federation, 2014), and an action plan for import substitution in the forest industry (Minpromtorg, 2015).

The Russian Prime Minister signed a decree in April 2015 outlining support for export producers (Government of the Russian Federation, 2015c). Also in 2015, the Ministry of Industry and Trade and the Ministry of Natural Resources and Environment developed a roadmap for the forest industry, including the "Strategy of Development of Forest Industry until 2030". An influential federal law in the field of forest relations (415-FZ; Rossiyskaya Gazeta, 2013) came into force in January 2016, building on the system for monitoring harvested timber to include administrative responsibilities for violations. The system addresses the reliability of data on forest users, lease agreements and purchase/sale agreements. A 2015 federal law simplified the procedure by which small and medium-sized businesses can participate in the purchase and sale of plantations for commercial timber harvesting (Rossiyskaya Gazeta, 2015). The new rules are designed to provide equal rights for small and large timber companies. Through such measures, the government expects to increase logging by 22 million m^3 per year.

In April 2016, the Russian government reviewed and approved the Paris Agreement – the main outcome of the 21st Conference of the Parties to the UNFCCC (Government of the Russian Federation, 2016).

Armenia and Kyrgyzstan officially joined the Eurasian Customs Union with Belarus, Kazakhstan and the Russian Federation in 2015 (Eurasian Economic Commission, 2015, 2016). The Eurasian Customs Union is designed to reduce barriers to the movement of goods, services, capital and labour, and it is expected to increase the trade in forest products among these countries.

Wood and wood products were in the media and policy spotlight in Belarus and Ukraine in early 2015. The log export ban introduced by Ukrainian Law 325-VIII of 9 April 2015 entered into force on 1 November 2015. The law prohibits Ukrainian exports of untreated wood from all tree species except pine for ten years (the export ban on pine will begin on 1 January 2017). The ban applies to all roundwood and to sawn timber with a thickness exceeding 70 mm and a moisture content greater than 22%. Belarus Decree 211 of May 2015, which banned the export of unprocessed wood, entered into force on 1 January 2016.

2.2.4 Due diligence and legal wood supply

2.2.4.1 EU Forest Law Enforcement, Governance and Trade Action Plan

The EU Forest Law Enforcement, Governance and Trade (FLEGT) Action Plan has been in force for 13 years. One of its key features is the negotiation of voluntary partnership agreements (VPAs) between the EU and tropical timber-supplying countries, which engage partner countries in the development of legality licensing systems for timber exported to the EU. Fifteen countries were negotiating or implementing VPAs in April 2016 (EU FLEGT Facility, 2016). On 21 April 2016, the EU and Indonesia issued a joint statement to announce that the licensing system for Indonesian timber was fully ready and that the first FLEGT-licensed timber would arrive in the EU later in 2016 (ITTO, 2016).

On 4 May 2016, the European Commission published an evaluation of the first 11 years of implementation (2003-2014) of the EU FLEGT Action Plan, drawing on a wide-ranging consultation process. The evaluation found that the Action Plan had contributed to improving forest governance globally and had probably helped reduce European demand for, and imports of, illegal timber. It also indicated, however, that the effectiveness of the Action Plan was hindered by difficulties encountered in producing FLEGT-licensed timber, and that the trade benefits of VPAs were yet to materialize. The evaluation concluded that while the overall aim of the Action Plan remains fully relevant, it needs to be adapted to address new challenges, particularly forest conversion, and more focus is needed on private-sector engagement and communication. The evaluation recommended a shift in geographical focus to non-VPA countries and greater emphasis on international coalitions to address illegal logging and trade (European Commission, 2016b).

On 19 February 2016, the European Commission published a review of the effectiveness of the EUTR, which was introduced in March 2013. The EUTR is part of the EU's policy on illegal logging and associated trade, which was defined in 2003 in the EU FLEGT Action Plan. By the end of 2015, 27 of the 28 EU member states had put in place national regimes for enforcing the EUTR. The only member state not to have implemented the necessary steps was Hungary, a relatively minor producer and importer of timber products. Norway (not an EU member) put the EUTR in place on 1 May 2015. Although most EU countries have taken the necessary legal steps to introduce the EUTR nationally, the review noted that penalties and enforcement measures varied widely (European Commission, 2016c).

There is evidence that the EUTR is contributing to significant changes in trade attitudes, structures and distribution networks. In a survey of timber-trading companies undertaken for the EUTR review in 2015, about one-third of European respondents reported changes in their sourcing of wood and wood products from tropical countries, and more than half were requesting suppliers to become certified. The survey also indicated that the EUTR imposed a cost and administrative burden on EU importers and their overseas suppliers and was encouraging the withdrawal from export markets of small and medium-sized enterprises in producer countries.

Government agencies in some EU member states have been subjecting timber products obtained from EU importers to microscopic, DNA and chemical analysis to identify – to the extent possible – the species and country of origin. Even though, in isolation, this form of analysis is rarely (if ever) sufficient to identify wood of illegal origin, importers are being sanctioned under the EUTR for their failure to correctly identify the species present in products on the grounds that it indicates inadequate due-diligence procedures. Importers have been put on notice that they must accurately identify the exact species composition of products, irrespective of the complexity of the supply chain or the underlying risks of illegal origin (ITTO IMM, 2016).

2.2.4.2 Lacey Act

The US Lacey Act, enacted in 1900, addresses trafficking in wildlife, fish and plants that have been illegally taken, possessed, transported or sold. After a series of amendments in 2008, the Act now requires that import declarations accompany certain plants and plant products, including a wide range of forest and wood products (Federal Register, 2015). In August 2015, the following product types were added to the phased-in enforcement schedule: casks, barrels, hunting knives and barbeque forks with wood handles, teak chairs, bent-wood seats and furniture, and rough wood blocks for the manufacture of smoking pipes (Federal Register, 2015). The US government has indicated that a process is underway to develop a proposal for establishing exceptions to the declaration requirement for composite materials and products containing minimal amounts of plant material (Federal Register, 2015).

In April 2015, the US Justice Department announced that it was seeking criminal charges under the Lacey Act against the company Lumber Liquidators, following a raid of its facilities (Associated Press, 2015). It was reported in October 2015 that Lumber Liquidators had made a settlement with the Environment and Natural Resources Division of the Department of Justice, pleading guilty to Lacey Act violations and agreeing to pay $10 million in fines, including contributions to conservation organizations (Koenig, 2015).

2.2.5 Bioenergy, biomass and biofuels

Data from the European Biomass Association show that bioenergy accounts for more than 60% of all renewable energy consumed in the EU28 and that about 70% of total bioenergy feedstock delivered in Europe originates in the forest sector (the rest comes from waste and agriculture). Bioenergy consumption almost doubled in the EU from 2000 to 2013, to 105.1 million tonnes of oil equivalent (Mtoe). It is set to grow by at least 33 Mtoe by 2020, according to the projections of EU member states, with biomass the only renewable energy source that provides solutions for all energy sectors (i.e. transport, power, and heating and cooling) (AEBIOM, 2016).

The European Commission introduced Directive 2015/1513 on 9 September 2015 to amend the sustainability rules for biofuels introduced by the Renewable Energy Directive (2009/28/EC (RED)) and the Fuel Quality Directive (2009/30/EC (FQD)). Only biofuels and bioliquids that comply with specified criteria can receive government support or be counted towards national renewable energy targets. The amended sustainability rules include provisions for addressing the impacts of land-use change, given that many biofuels are produced from crops grown on existing agricultural land. EU member states are required to bring into force, by 10 September 2017, the necessary laws and regulations to ensure compliance at the national level.

The amended sustainability rules state that biofuels arising from food crops can contribute no more than 7% towards the 2020 renewable energy targets of member states. The rules also allocate more credit for achieving the targets using biofuels that do not place additional demands on land. All forms of woody biomass except sawlogs and veneer logs are given extra credit alongside biomass and biofuels from algae and various forms of municipal, industrial and agricultural waste. The amended rules also state that installations that started producing biofuels on or after 6 October 2015 must achieve a saving of greenhouse gas emissions of at least 60% compared with fossil fuels. The percentage saving must be at least 35% for installations that started producing biofuels before 6 October 2015. This latter target must be met before December 2017; thereafter, the minimum saving increases to 50% (European Commission, 2016d).

Despite the new rules, 120 civil-society organizations and networks worldwide published a declaration in February 2016 calling for bioenergy to be totally excluded from the next EU Renewable Energy Directive. The central premise of the declaration is that bioenergy should not be classed as renewable energy on the grounds that demand exceeds the rate of replenishment and that biomass carbon emissions are underestimated (World Rainforest Movement, 2016). This view contradicts the scientific literature, however, which suggests that bioenergy could play a more prominent role in reducing greenhouse gas emissions if appropriate policies and safeguards are in place (e.g. Mendes Souza et al., 2015).

The Sustainable Biomass Partnership (SBP), created in 2013 to continue the work of the former Initiative of Wood Pellet Buyers, launched the Biomass Assurance Framework Standards (version 1.0) in March 2015. The SBP approved the first two certification bodies in September 2015, as well as the first certified biomass producer in the US, followed soon (in October 2015) by the first certified biomass producer in Europe (Sustainable Biomass Partnership, 2016a). By March 2016, the SBP had grown to include 16 certified organizations, including the first biomass trader, and encompass eight countries – Belarus, Denmark, Estonia, Latvia, Lithuania, Poland, Portugal and the US. An additional 70 organizations, including producers and traders, have applied for SBP certification (Sustainable Biomass Partnership, 2016b).

2.3 CERTIFIED FOREST AREA

The two major certification schemes – the FSC and the PEFC – reported a combined global total of 462 million hectares of certified forestlands in 2016 (nominally to May); (FSC, 2016a; PEFC, 2016a); (graph 2.3.1). This was an increase of 15.8 million hectares (3.5%) compared with the previous 12 months, significantly higher than the increase of about 6.2 million hectares (1.4%) in the 12 months before that. An estimated 29.5 million hectares of forestland is certified under both schemes; correcting for this overlap, the total forest area certified globally is 432.5 million hectares (11% of the global forest area).

The estimated overlap of forestlands certified under both the FSC and PEFC-endorsed schemes was up by some 20 million hectares compared with previous years, when overlaps amounted to between 5 million and 8 million hectares (the large adjustment is due to improved data availability). Ninety-five percent of the double-certified forest area is in North America (24 million hectares) and Europe (4 million hectares). About 13 million hectares of forestland in Canada is certified by both the Sustainable Forestry Initiative (SFI, a PEFC-endorsed scheme) and the FSC; in the US, the area certified by the two schemes amounts to just over 7 million hectares. The PEFC-endorsed American Tree Farm System (ATFS) overlaps with the FSC on just over 1 million hectares in the US. In Canada, the PEFC-endorsed Canadian Standards Association (CSA) certification scheme overlaps with FSC-certified forests on just more than 3 million hectares. It should be noted that there is ongoing research to better assess forestland with double certification and the results should be available from PEFC and FSC in late 2016.

GRAPH 2.3.1

Cumulative forest area certified by major certification schemes, 2006-2016

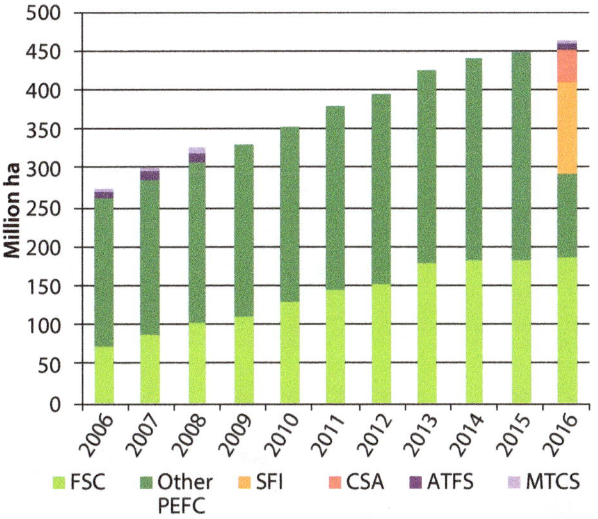

Notes: **FSC and SFI data as of May 2016; PEFC, CSA, ATFS and MTCS data as of and including March 2016. Data for systems endorsed by the PEFC (the ATFS, the CSA, the MTCS, and the SFI) are included in the PEFC data after the date of endorsement. Data for 2016 differentiate by the major systems endorsed by the PEFC. Overlaps due to double certification are not taken into account.**

Source: **FSC, 2016a; PEFC, 2016a; SFI, 2016a; CSA, 2016b; ATFS, 2016; MTCS, 2016.**

2.3.1 Internal developments in certification schemes

The process to develop the FSC's international generic indicators (IGIs) was completed when the FSC's board of directors gave its final approval in March 2015, making the IGIs available to be used in the development of national and subregional FSC standards (FSC, 2015a). In March 2016, the FSC announced that Portugal was the first national initiative to use the IGIs in the revision of its national standard and that the scheme in that country had duly transferred to the fifth version of the FSC Principles and Criteria (FSC, 2016b).

The FSC's board of directors decided to end the FSC's disassociation from Dalhoff Larsen and Hornewman (DLH) in January 2016 on a probationary basis (FSC, 2016c). The disassociation was instigated in February 2015 in light of evidence that DLH had been involved in illegal timber trade in Liberia (FSC, 2015b). The FSC accepted two new complaints cases under its policy on association in 2015 (FSC, 2015c). One of these, in Fiji, relates to national legislation that is incompatible with the International Labour Organization's core conventions on freedom of association and collective bargaining (FSC, 2015d). The other case pertains to complaints of illegal logging in Romania (FSC, 2015e).

In North America, the SFI and its partners set a Guinness World Record by planting 202,935 trees in an hour. The effort involved 29 teams of 25-100 people planting trees in locations from New York City, US, to Vancouver Island, Canada (SFI, 2015). A fourth edition of Canada's National Standard on Sustainable Forest Management, CAN/CSA-Z809, was published in 2016 (CSA, 2016a).

The PEFC's effort to support certification in Africa moved forward in 2015 with the completion of the first pre-audit of the PEFC-endorsed national forest certification system in Gabon (PEFC, 2015a). The first PEFC-certified forests in Indonesia were recognized in 2015, comprising more than 600,000 hectares managed by the companies Asia Pacific Resources International and Asia Pulp & Paper (PEFC, 2015b). In the past year, India and Hungary became the 40th and 41st national-level members of the PEFC – through the Network for Certification Conservation of Forests (India) and the Hungarian Forest Certification Non-profit Ltd – thus continuing the development of national forest certification systems in those two countries (PEFC, 2015c; PEFC, 2016b). The New Zealand Forest Certification Scheme – based on the New Zealand Standard for Sustainable Forest Management (NZS AS 4708:2014) – achieved PEFC endorsement in December 2015 (PEFC, 2015d).

2.3.2 Regional aspects

The PEFC reported a certified forest area of 275.3 million hectares in 34 countries (two countries more than in the previous survey), as of March 2016; this was up by 12.1 million hectares (4.6%), year-on-year.

The FSC reported a certified forest area of 183.3 million hectares in April 2016 in the same 81 countries as in the previous 12 months. A large part of the FSC-certified forest is in Europe (88.7 million hectares) and North America (67 million hectares); 12.9 million hectares are FSC-certified in Latin America and the Caribbean.

The PEFC endorsed the certification of 5.6 million hectares in China and 1.1 million hectares in Indonesia in 2015; previously it had endorsed the Malaysian Timber Certification Scheme (MTCS), encompassing the certification of 3.9 million hectares of forest in Malaysia. With the addition of forest in China and Indonesia, Asia now accounts for the third-largest share of PEFC-certified forest (10.6 million hectares) of any region, after North America (163.7 million hectares) and Europe (85.4 million hectares).

The PEFC has endorsed the ATFS, CSA and SFI standards, which combined represent about 60% of PEFC-endorsed certified forest globally. SFI forest management certification represents about 40% of PEFC certification globally; its certified forest area has more than doubled since 2007, to 115 million hectares (as of March 2016).

Eighty-seven percent of the global certified forest area is in the Northern Hemisphere (graph 2.3.2). Forty-eight percent (207 million hectares) of the certified forest area globally is in North America, 25% (107 million hectares) is in western Europe, and 14% (63 million hectares) is in the CIS. Combined, Latin America, Africa, Asia and Oceania account for only 13%

(56 million hectares) of certified forests worldwide, even though these regions contain 60% (2.4 billion hectares) of the global forest estate.

GRAPH 2.3.2

Share of certified forest management area, by region, 2016

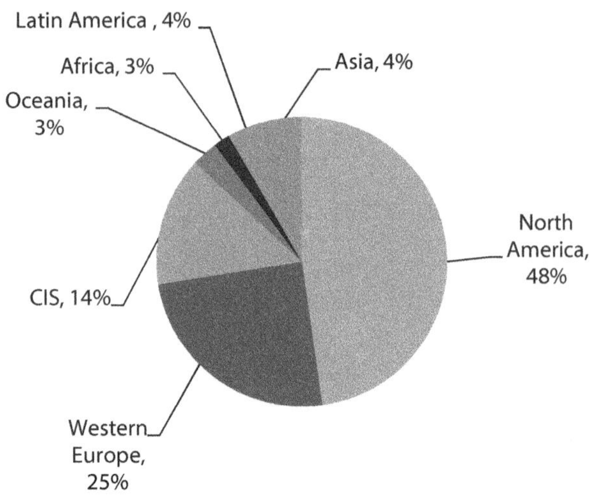

Notes: Data for systems endorsed by the PEFC (the ATFS, the CSA, the MTCS and the SFI) are included in the PEFC data after the date of endorsement. The data take into account an estimated overlap of 29.5 million hectares (as of May 2016) due to double certification. FSC data are as of April 2016; PEFC data are as of March 2016.
Sources: FAO, 2010, 2015; FSC, 2016a; PEFC, 2016a; SFI, 2016a,b.

Table 2.3.1 and graph 2.3.3 shows the change in certified forest area under the FSC and the PEFC in seven regions globally in the period 2014-2016. The PEFC lost about 3.5% (2.6 million hectares) in western Europe and 11.5% (1.3 million hectares) in the CIS, respectively, over the period, but it increased its certified area in North America by 6% (9.5 million hectares) and more than doubled its share in Asia, to 6 million hectares. To date, the PEFC has not certified any forests in Africa; in contrast, the FSC increased its certified forest area on that continent by 19% (1.2 million hectares) over the period. The FSC also increased its certified forest area in Oceania (by 0.1 million hectares, up 4%), the CIS (by 1.3 million hectares, up 2.5%) and North America (by 1.4 million hectares, up 2%). Overall, the FSC-certified forest area increased in the period in all regions except Asia, where it declined by 9% (0.8 million hectares).

TABLE 2.3.1

Estimated global and regional supply of roundwood from certified resources, 2014-2016

	Total forest area (million ha)	*Certified forest area (million ha)*			*Certified forest area (%)*			*Estimated industrial roundwood from certified forests (million m³)*			*Estimated proportion of total roundwood production from certified forests (%)*		
		2014	*2015*	*2016*	*2014*	*2015*	*2016*	*2014*	*2015*	*2016*	*2014*	*2015*	*2016*
North America	614.2	221.3	217.3	206.8	36.0	35.4	33.7	250.5	245.9	234.0	14.1	13.9	13.2
Western Europe	168.1	106.6	109.6	106.8	63.4	65.2	63.6	251.1	258.1	251.7	14.2	14.6	14.2
CIS	836.9	55.5	62.9	62.9	6.6	7.5	7.5	10.6	12.0	12.0	0.6	0.7	0.7
Oceania	191.4	12.6	12.5	12.6	6.6	6.5	6.6	3.6	3.6	3.6	0.2	0.2	0.2
Africa	674.4	6.4	6.5	7.8	1.0	1.0	1.2	1.9	2.0	2.3	0.1	0.1	0.1
Latin America	955.6	16.3	17.1	17.8	1.7	1.8	1.9	1.2	1.3	1.3	0.1	0.1	0.1
Asia	592.5	14.1	13.1	18.3	2.4	2.2	3.1	4.5	4.2	5.9	0.3	0.2	0.3
World total	4,033.1	432.8	439.0	432.8	10.7	10.9	10.7	523.5	527.1	510.9	29.6	29.8	28.8

Notes: Estimates of forest area (excluding "other wooded land") and industrial roundwood production from certified forests are based on data in FAO (2010). The annual roundwood production in "forests available for wood supply" in a given region or subregion is multiplied by the percentage of that region or subregion's certified forest area (i.e. it is assumed that the removals of industrial roundwood from each hectare of certified forest is the same as the average for all forest available for wood supply). Not all certified roundwood is sold with a label. 2014, 2015 and 2016 represent May of the prior year to May of the current year. "World" is not a simple total of the regions and subregions. The data shown take into account estimated overlaps of 29.5 million hectares (as of May 2016) and 7.5 million hectares (as of May 2014 and May 2015) due to double certification. Information is valid as of May 2016. FSC data are as of April 2016; PEFC data are as of March 2016.

Sources: FAO, 2010; FSC, 2016a; PEFC, 2016a; SFI, 2016a, b; authors' compilation.

GRAPH 2.3.3

Certified forest area by region and certification scheme, 2014-2016

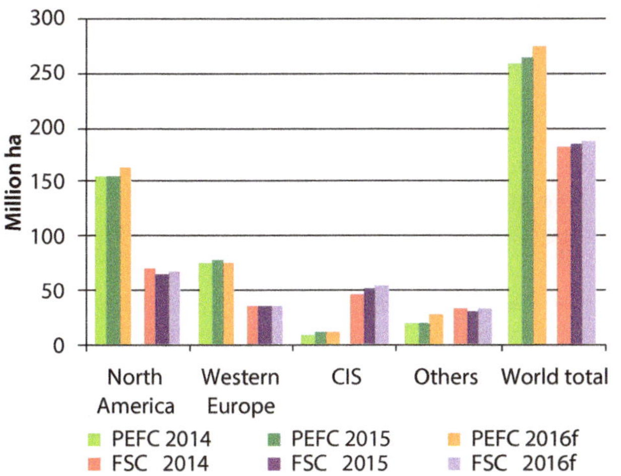

Notes: Data for systems endorsed by the PEFC (the ATFS, the CSA, the MTCS and the SFI) are included in the PEFC data after the date of endorsement. The data do not take into account overlaps due to double certification. FSC data are as of April 2016; PEFC data are as of March 2016.

Sources: FSC, 2016a; PEFC, 2016a.

2.3.3 Certified forest production

Table 2.3.1 shows a reduction in estimated roundwood production in certified forests in the 12 months to May 2016, from 527 million m^3 to 511 million m^3, due largely to the detection of overlaps in certified areas between the FSC and PEFC schemes. Most of the reduction was in North America, where estimates of certified roundwood production decreased by 12 million m^3, even though the FSC and the PEFC both reported increases in their areas of certified forest in the subregion. North America is estimated to produce about 234 million m^3 of certified roundwood, which is 13.2% of global roundwood production. For the third year in a row, the production of certified roundwood as a proportion of global roundwood production was higher in Europe than in North America, at 14.2% (252 million m^3) in 2016. One reason for the higher certified roundwood production in Europe – despite less growth in the certified forest area – is the comparatively small area of forests certified by both the FSC and PEFC schemes (4 million hectares, compared with 24 million hectares in North America). Africa and Asia both increased their estimated certified roundwood production in the 12 months to May 2016, by 300,000 m^3 and 1.7 million m^3, respectively.

2.3.4 Chain-of-custody certification

For both the FSC and the PEFC, the most important markets for chain-of-custody (CoC)-certified products are Asia, Europe and North America. In the 12 months to May 2016, the FSC issued 30,380 CoC certificates in 117 countries (five more countries than in the previous 12-month period), and the PEFC issued 10,853 CoC in 53 countries (seven more countries than in the previous 12 months). Graph 2.3.4 shows that growth in the number of CoC certificates after 2012 was significantly higher for the FSC scheme than for the PEFC scheme, although the pace of growth in FSC certificates also slowed after 2014. Overall, the growth in the number of CoC certificates declined considerably between 2008 and 2016, from 46% to 4%.

GRAPH 2.3.4

Number of chain-of-custody certificates issued globally by the FSC and the PEFC, 2008-2016

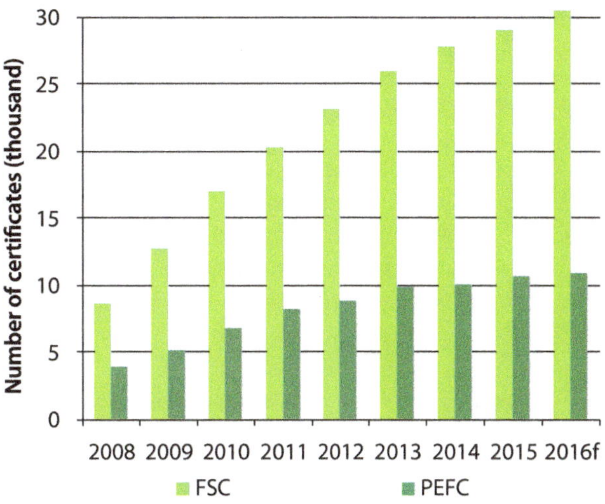

Notes: Numbers denote CoC certificates irrespective of the size of the individual companies or of production or trade volumes. FSC data are as of April 2016; PEFC data are as of March 2016.

Sources: FSC, 2016a; PEFC 2016a.

2.4 CARBON-RELATED

2.4.1 Climate change and carbon markets

The 21st Conference of the Parties (COP 21) to the UNFCCC in Paris, France, in 2015 produced the Paris Agreement, which aims to accelerate actions to mitigate climate change (UNFCCC, 2015).

Key features of the Paris Agreement include: a goal of limiting global warming to less than 2 degrees Celsius above pre-industrial levels; a global transition away from fossil fuels, with 189 national climate plans covering 98% of all emissions; and enhanced transparency, accountability and tracking (European Commission, 2016e).

UNFCCC COP 21 raised the political profile of forestry (the only economic sector referenced explicitly in the Paris Agreement); it signalled that cutting emissions from deforestation and promoting sustainable forestry are now recognized globally as among the most efficient ways to address climate change.

The Paris Agreement is built on the commitment of signatories to deliver against "intended nationally determined contributions" (INDCs) to reduce greenhouse gas emissions. By allowing countries to voluntarily declare their own commitments, discussions in Paris sidestepped the political conflict created in earlier negotiations, which sought to allocate specific targets for emission reductions to individual countries.

The key element in the Paris Agreement is the long-term mitigation goal in Article 4.1, which enshrines the concept of reducing global emissions to zero in the second half of the twenty-first century. This would be achieved through "a balance between anthropogenic sources and removals by sinks of greenhouse gases in the second half of this century". Article 5 anchors forests in the Agreement by requiring parties to take action to conserve and enhance natural ecosystems, including forests. Parties are encouraged to implement and support the existing REDD+ framework[3] through both results-based payments and alternative policy approaches. The Agreement does not, however, introduce any new financial commitments, pledges or channels for REDD+, beyond the general provisions for climate finance set out in Article 9.

Article 6.4 establishes a sustainable development mechanism (SDM), which will operate as an offset mechanism by which parties (and private entities) can sell emissions reductions to another party. The new trading provisions are open to both developed and developing countries, meaning that – similarly to the Kyoto Protocol's Joint Implementation programme – any signatories can be buyers or sellers of emission units, now called "internationally transferred mitigation outcomes" (ITMOs). Although the Paris Agreement doesn't mention "markets" explicitly, Parties can voluntarily use ITMOs to help meet the reduction targets set out in their INDCs.

Although still open to negotiation, forest carbon initiatives are likely to be eligible for ITMO trading. To meet requirements, emission reductions must be "real, measurable, and long-term" and verified by "designated operational entities" within a framework supervised by a body to be designated by the 196 UNFCCC delegations (likely to be the UNFCCC itself). The mechanism is open to both public and private entities (UNFCCC, 2015).

The SDM is expected to replace the existing UN offset scheme, the Clean Development Mechanism (CDM), before 2020 when more detailed rules and procedures have been finalised. An important issue for the forest sector is to ensure that the SDM is more relevant to forest carbon initiatives than the CDM. As of July 2016, only 66 of 8073 registered CDM projects were forestry related and these projects accounted for only 0.7% of all certified emission reductions issued (UNEP DTU, 2016).

The EU and its member states have committed to a binding minimum target of a 40% domestic reduction in greenhouse gas emissions from 1990 levels by 2030, as well as to renewable energy and energy efficiency targets. After UNFCCC COP 21, the EU decided to maintain its established climate targets for 2030. Signatory countries to the Paris Agreement need to establish policy actions to achieve commitments, and the European Commission has initiated this process with a proposal to revise the EU Emissions Trading System. Additional legislative proposals to implement the 2030 targets are expected in 2016 and 2017 (European Commission, 2016f).

The EU Emissions Trading System is the dominant world carbon market, but it continues to be affected by oversupply and low carbon prices. A surplus of allowances has built up in the system since 2009; the build-up has slowed in recent years but the overall surplus is not expected to decline significantly before 2020 from the current level of around 2 billion allowances (European Commission, 2016c). Sufficient surplus credits were available in 2015 to cover 90% of emissions by heavy emitters; there were 1.7 billion surplus allowances in the market in that year, down from 2.1 billion tonnes in 2014. The EU carbon price continues to be in the range of €5-6 per tonne, significantly lower than the €30 per tonne once predicted – and which is widely speculated to be the level at which the market would contribute effectively to reducing emissions (Taylor, 2016).

2.4.2 Green building

According to the latest edition of the Global Sustainable Buildings Index (Larroque et al., 2016), which examines green building practices in 25 countries in Asia and the Pacific, Europe, the Middle East and the Americas, European countries continue to lead such practices. The Index classes France, Germany, the Netherlands and the UK as "Category 1" countries, with the highest combined scores against criteria addressing green certification, regulation, retrofit incentives, targets for carbon dioxide emissions and energy, and financing. Three European countries – Belgium, Italy and Spain – as well as Canada, Singapore and the US are classed as "Category 2".

Nevertheless, the European Commission's 2014 Communication on Resource Efficiency Opportunities in the Building Sector identified the need for a common EU framework of indicators for the assessment of the environmental performance of buildings. The Commission launched a three-year project to develop this approach in 2015. The first output of the project, released in December 2015, was a working paper identifying the following six high-level objectives as a basis for the forthcoming EU green building policy: 1) greenhouse gas emissions from building life-cycle energy use; 2) resource-efficient material life cycles; 3) the efficient use of water resources; 4) healthy and comfortable spaces; 5) resilience to climate change; and 6) optimized life-cycle cost and value (Dodd et al., 2015).

Meanwhile, the EU is focusing efforts on improving the energy efficiency of existing buildings, which are estimated to account for 36% of the EU's total carbon emissions. The EU Energy Efficiency Directive requires member states to submit long-term strategies by 30 April 2017 for mobilizing investments in the renovation of existing residential and commercial properties to high energy-efficiency standards. In response to concerns that these strategies may be insufficient to meet long-term climate targets, 13 "green building councils" across Europe launched the BUILD UPON project in March 2016, with financial support from the EU's Horizon 2020 programme. The project aims to engage at least 1,000 key stakeholders – from governments and businesses to non-governmental organizations and householders – in 13 countries to help deliver strategies for renovating existing buildings by the 30 April 2017 deadline.[4]

[3] REDD+ is the term given to the concept of reducing emissions from deforestation and forest degradation, plus the sustainable management of forests and the conservation and enhancement of forest carbon stocks.

[4] http://buildupon.eu

Sources: E. O'Driscoll, 2016.

LEED ("Leadership in Energy and Environmental Design") is a building rating and certification programme developed by the US Green Building Council (USGBC). Some stakeholders have been critical of the programme for its treatment of forest products, including its limited recognition of forest certification programmes. In April 2016, the USGBC announced a new credit in the LEED programme designed to address illegal wood and promote the use of verified-legal wood in buildings (Holowka, 2016). The new pilot Alternative Compliance Path (ACP) credit categorizes the various forest certification standards based on the ASTM D7612-10 standard for "Categorizing Wood and Wood-based Products According to their Fiber Sources" and provides a tiered structure for evaluating wood products that are legal (e.g. non-controversial), responsibly sourced, and certified. The pilot ACP credit is applicable to the LEED 2009 and LEED v4 systems (Long, 2016). The PEFC, and PEFC-endorsed programmes such as the ATFS, the CSA and the SFI, welcomed the change as an opportunity to increase the recognition of products certified to their standards within the LEED programme (PEFC, 2016c). The FSC's certification programme is also recognized in the pilot ACP credit (FSC, 2016d).

Interest in tall wooden buildings and mass timber construction continues to grow. A barrier to innovation is that building codes in many jurisdictions limit wood structures to no more than 4-6 stories in height, even though technological innovations mean that wood construction can achieve 40 stories or more (Bowyer et al., 2016). At least 21 wood structures more than six stories in height have been built worldwide since 2009; many of these have been enabled by wooden-building initiatives and supportive policies (Bowyer et al., 2016).

2.4.3 Environmental product declarations

The pilot phase of an initiative to develop a harmonized framework for product environmental footprints (PEFs) as part of the EU's "Single Market for Green Products" is due to be completed in 2016. Concerns have been expressed in the European construction sector that the PEF framework should take greater account of the pioneering work already carried out to develop a harmonized framework for environmental product declarations (EPDs) for building products, in line with the EN15804 standard. [5] ECO Platform, which represents all major European EPD programme operators in the construction sector, issued a position paper in January 2016 expressing concern about the "seemingly rapid introduction of the PEF methodology" and suggesting that rather than a preference for PEFs there should be "a convergence between relevant ISO standards, EN15804 and the PEF methodology for construction products" (ECO Platform, 2016). More than 2,000 EN15804-conformant EPDs had been issued by the end of 2015 for construction-sector products, of which around 200 were EPDs issued in accordance with harmonized procedures developed within the ECO Platform.

The EU's Action Plan for the Circular Economy, launched in December 2015, encompasses a wide range of actions to promote eco-design based on the reparability, durability and recyclability of products; energy efficiency; improved labelling that takes circular economy principles into account; circular economy criteria in green public procurement; improved waste management; and developing markets for secondary raw materials. The implementation of the Action Plan for the Circular Economy is supported by €650 million from the EU's Horizon 2020 fund for research and innovation and €5.5 billion in structural funds for waste management, in addition to national-level investments. The Plan establishes the following EU targets: recycling 65% of municipal waste by 2030; recycling 75% of packaging waste by 2030; reducing landfill to a maximum of 10% of all waste by 2030; and a total ban on landfilling of separately collected waste.

The BioPreferred programme of the US Department of Agriculture (USDA) was launched five years ago; today, about 2,500 products carry the USDA Certified Biobased Product label, including more than 100 product categories (USDA, 2016a). A number of forest products are included in the BioPreferred catalogue, including lumber, papers, packaging and tissues (USDA, 2016b).

The UN General Assembly approved the 17 Sustainable Development Goals (SDGs) – with their 169 targets – on 25 September 2015, replacing the eight Millennium Development Goals (MDGs) adopted in 2000. Unlike for the MDGs, both developed and developing countries have committed to attaining the SDGs. SDG 15 sets a target for ending deforestation and achieving the conservation, restoration and sustainable use of forests and other ecosystems by 2020, in line with obligations under international agreements. SDG 15 calls for the mobilization of significant resources from all sources and at all levels to finance sustainable forest management and to provide developing countries with adequate incentives to advance such management.

2.5 CONCLUSION

Significant developments affecting markets for forest products continue to occur in trade policies, environmental and climate initiatives, and voluntary programmes. In many ways, these developments have been in a positive direction; uncertainties remain on trade and climate policies, however, and there are mixed trends in certification programmes. Greater alignment of these diverse efforts is needed to realize the full benefits of forests and forest products.

[5] EN15805 "Sustainability of construction works – Environmental product declarations – Core rules for the product category of construction products".

2.6 REFERENCES

AEBIOM. 2016. Bioenergy accounts for 60% of EU renewable energy. News report on European Biomass Association (AEBIOM) Statistical Report 2015. October 27, 2015.

Associated Press. 2015. Justice Department now seeking criminal charges against Lumber Liquidators over imports. 29 April 2015. Available at: www.foxbusiness.com/markets/2015/04/29/justice-department-now-seeking-criminal-charges-against-lumber-liquidators-over

ATFS. 2016. ATFS website. American Tree Farm System (ATFS). Available at: www.treefarmsystem.org/certification-american-tree-farm-system

Bowyer, J., Bratkovich, S., Howe, J., Fernholz, K., Frank, M., Hanessian, S., Groot, H. and Pepke, E. 2016. Modern tall wood buildings: opportunities for innovation. Available at: www.dovetailinc.org/report_pdfs/2016/dovetailtallwoodbuildings0116.pdf

Canadian Boreal Forest Agreement. 2015. Canadian Boreal Forest Agreement. Available at: http://cbfa-efbc.ca/agreement

Christensen, N. 2015. Branching out: preparing for life without a softwood lumber agreement. Canada West Foundation. Available at: http://cwf.ca/research/publications/branching-out-preparing-for-life-without-a-softwood-lumber-agreement

CSA. 2016a. Z809-16 Sustainable forest management. Canadian Standards Association (CSA). Available at: http://shop.csa.ca/en/canada/sustainable-forest-management/z809-16/invt/27017442016

CSA 2016b. Canadian sustainable forest management. Canadian Standards Association (CSA). Available at: www.csasfmforests.ca

Dodd, N., Donatello, S., Garbarino, E. and Gama-Caldas, M. 2015. Identifying macro-objectives for the life cycle environmental performance and resource efficiency of EU building. JRC Science and Policy reports. Working Paper 1. Institute for Prospective Technological Studies, Joint Research Centre, European Commission.

ECO Platform. 2016. Position paper on the introduction of an EU-initiated PEF-methodology into the construction sector. Available at: www.eco-platform.org/files/download/Documents/ECOPlatform_position_paper_PEF_2016-01-19.pdf

Eurasian Economic Commission. 2015. Armenia is now in the Eurasian Economic Union. Available at: www.eurasiancommission.org/en/nae/news/Pages/02-01-2015-1.aspx

Eurasian Economic Commission. 2016. Kyrgyzstan acceded to the Eurasian Economic Union. Available at: www.eurasiancommission.org/en/nae/news/Pages/12-08-2015-1.aspx

European Commission. 2016a. The Commission publishes further TTIP documents in ongoing transparency commitment. 21 March 2016 (accessed 8 April 2016). Available at: http://trade.ec.europa.eu/doclib/press/index.cfm?id=1477

European Commission. 2016b. Evaluation of the EU FLEGT Action Plan (Forest Law Enforcement Governance and Trade) 2004-2014. April 27, 2016. Available at: http://ec.europa.eu/europeaid/file/48831/download_en?token=CDg-_Q_P

European Commission. 2016c. EUTR Implementation Report. February 27, 2016. Available at: http://ec.europa.eu/environment/forests/eutr_report.htm

European Commission. 2016d. EU Directive 2015/1513 amending Directive 98/70/EC relating to the quality of petrol and diesel fuels and amending Directive 2009/28/EC on the promotion of the use of energy from renewable sources. September 9, 2016. Available at: http://eur-lex.europa.eu/legal-content/EN/TXT/PDF/?uri=CELEX:32015L1513&from=EN.

European Commission. 2016e. Communication from the Commission to the European Parliament and the Council. The Road from Paris: assessing the implications of the Paris Agreement and accompanying the proposal for a Council decision on the signing, on behalf of the European Union, of the Paris Agreement adopted under the United Nations Framework Convention on Climate Change. 2 March 2016. COM(2016) 110 final.

European Commission. 2016f. The EU Emissions Trading System. Available at: http://ec.europa.eu/clima/policies/ets/index_en.htm

EU FLEGT Facility. 2016. VPAs for SDGs: how voluntary partnership agreements contribute to global development goals. Factsheet. Available at: www.euflegt.efi.int/documents/10180/301892/VPAs+for+SDGs/2ba6b3ae-400d-45a3-8d40-69e2bce90900

FAO 2010. Global forest resources assessment 2010. Main report. Rome. Available at: www.fao.org/forestry/fra/fra2010/en

FAO 2015. Global forest resources assessment 2015. Rome. Available at: www.fao.org/forest-resources-assessment/current-assessment/en/

Federal Register. 2015. Implementation of revised Lacey Act provisions. 6 February 2015. Volume 80, No. 25. Available at: www.aphis.usda.gov/plant_health/lacey_act/downloads/APHIS-2008-0119-0288.pdf

FSC. 2015a. FSC board approves international generic indicators. 27 March 2015. Forest Stewardship Council (FSC). Available at: https://ic.fsc.org/newsroom.9.1104.htm

FSC. 2015b. Forest Stewardship Council disassociates from the DLH Group. 12 February 2015. Forest Stewardship Council (FSC). Available at: https://ic.fsc.org/fsc-press-releases.638.1065.htm

FSC. 2015c. Dispute resolution program update: December 2015. Two new FSC Policy for Association cases were accepted. 11 December 2015. Forest Stewardship Council (FSC). Available at: https://ic.fsc.org/en/news/id/1359

FSC. 2015d. Building and Wood Workers International and the International Trade Union Congress vs. Fiji Pine, Tropik Wood Industries, and Tropik Wood Products. Forest Stewardship Council (FSC). Available at: https://ic.fsc.org/en/stakeholders/dispute-resolution/current-cases/fiji-pine-tropik-wood-industries-and-tropik-wood-products

FSC. 2015e. WWF Germany vs. Holzindustrie Schweighofer. Forest Stewardship Council (FSC). Available at: https://ic.fsc.org/en/stakeholders/dispute-resolution/current-cases/holzindustrie-schweighofer-hs

FSC 2016a. FSC facts and figures April 2016. Forest Stewardship Council (FSC). Available at: https://ic.fsc.org/en/facts-figures

FSC. 2016b. Portugal becomes first country to transfer its national Forest Stewardship Standard to the FSC Principles and Criteria Version 5. 22 March 2016. Forest Stewardship Council (FSC). Available at: https://ic.fsc.org/en/news/id/1454

FSC. 2016c. FSC ends disassociation with Dalhoff Larsen and Horneman. 18 January 2016. Forest Stewardship Council (FSC). Available at: https://ic.fsc.org/en/news/id/1394

FSC. 2016d. Pilot testing legality for forest products in LEED. 14 April 2016. Forest Stewardship Council (FSC). https://us.fsc.org/en-us/newsroom/newsletter/id/934

GAC. 2016a. Joint statement by European Commissioner for Trade and Canada's Minister of International Trade on Canada-EU trade agreement. 29 February 2016. Global Affairs Canada (GAC). Available at: http://news.gc.ca/web/article-en.do?mthd=index&crtr.page=1&nid=1036759

GAC. 2016b. GAC, 2016b. Global Affairs Canada, Canada-European Union: Comprehensive Economic and Trade Agreement (CETA). Web page available at: http://www.international.gc.ca/trade-agreements-accords-commerciaux/agr-acc/ceta-aecg/benefits-avantages/forest-foret.aspx?lang=eng

Government of the Russian Federation. 2014. Decree of the Government of the Russian Federation No. 328 of 15 April 2014. Available at: http://government.ru/docs/11912

Government of the Russian Federation. 2015a. Decrees of the Government of the Russian Federation No. 892 of 27 August 2015; No. 861 of 19 August 2015; and No. 838 of 13 August 2015. Available at: http://government.ru/media/files/rUFPqtrlxczUvv5vL9AAKCRU5O9B1BHs.pdf; http://government.consultant.ru/documents/3704151; http://government.ru/media/files/Uz6dK9P1PDUAOPf9rBVGcHfj7hKEdz8C.pdf

Government of the Russian Federation. 2015b. Government Resolution No. 98-r of 27 January 2015. Available at: http://government.ru/docs/16639

Government of the Russian Federation. 2015c. Decree of the Government of the Russian Federation No. 593-p of 3 April 2015. Available at: http://government.ru/docs/17595/

Government of the Russian Federation. 2016. Decree of the Government of the Russian Federation No. 670-p of 14 April 2016. Available at: http://government.ru/en/docs/22726

Holowka, T. 2016. Illegal wood and the role of LEED. 21 April 2016. Available at: www.usgbc.org/articles/illegal-wood-and-role-leed-0

ITTO. 2016. Indonesia and EU ready to move on FLEGT licensing. News release 28 April 2016. International Tropical Timber Organization (ITTO). Available at: www.itto.int/news_releases/id=4794

ITTO IMM. 2016. FLEGT Independent Market Monitoring Annual Summary Report 2015. Section on EUTR: Extent of private sector compliance. Draft version. International Tropical Timber Organization (ITTO) Independent Market Monitoring (IMM). 25 May 2016.

Koenig, K. 2015. Lumber Liquidators settles Lacey Act violations for $10 million. 7 October 2015. Available at: www.woodworkingnetwork.com/news/woodworking-industry-news/lumber-liquidators-settles-lacey-act-violations-10-million

Larroque, J., Lechien, D., Farnell, B., Rubin, J. and Co, A., eds. 2016. Global Sustainable Buildings Index. 2nd edition. Baker and McKenzie. Available at: http://f.datasrvr.com/fr1/016/40926/Global_Sustainable_Buildings_Index_2nd_edition_FULL_VERSION.pdf

Long, M. 2016. USGBC announces new LEED pilot ACP designed to help eliminate irresponsibly sourced materials – like illegal wood – from the building material supply chain. 5 April 2016. Available at: www.usgbc.org/articles/usgbc-announces-new-leed-pilot-acp-designed-help-eliminate-irresponsibly-sourced-materials%E2%80%94

Mendes Souza, G., Victoria, R.L., Joly, C.A. and Verdade, L.M., eds. 2015. Bioenergy & sustainability: bridging the gaps. Scientific Committee on Problems of the Environment, 72. Available at: http://rsb.org/pdfs/reports/Bioenergy%20and%20Sustainability%20-%20Bridging%20the%20Gaps.pdf

Minpromtorg. 2015. Order of Ministry of Industry and Trade of the Russian Federation (Minpromtorg). Available at: http://minpromtorg.gov.ru/common/upload/files/docs/657.pdf

MTCS. 2016. Malaysian Timber Certification Scheme [MTCS]. Available at: http://mtc.com.my/timber-certification-in-malaysia/

PEFC. 2015a. Gabon takes next step towards PEFC-certified forests. 18 November 2015. Programme for the Endorsement of Forest Certification (PEFC). Available at: http://pefc.org/news-a-media/general-sfm-news/1974-gabon-takes-next-step-towards-pefc-certified-forests

PEFC. 2015b. Indonesia celebrates its first PEFC-certified forests. 9 June 2015. x Programme for the Endorsement of Forest Certification (PEFC). Available at: http://pefc.org/news-a-media/general-sfm-news/1873-indonesia-celebrates-its-first-pefc-certified-forests

PEFC. 2015c. PEFC milestone: we welcome our 40th member. 28 September 2015. Programme for the Endorsement of Forest Certification (PEFC). Available at: http://pefc.org/news-a-media/general-sfm-news/1940-pefc-milestone-we-welcome-our-40th-member

PEFC. 2015d. New Zealand forest owners to benefit from international recognition. 10 December 2015. Programme for the Endorsement of Forest Certification (PEFC). Available at: http://pefc.org/news-a-media/general-sfm-news/1987-new-zealand-forest-owners-to-benefit-from-international-recognition

PEFC 2016a. PEFC global statistics: SFM & CoC certification. March 2016. Programme for the Endorsement of Forest Certification (PEFC). Available at: www.pefc.org/images/documents/PEFC_Global_Certificates_-_March_2016.pdf

PEFC. 2016b. Hungary becomes latest PEFC member. 20 January 2016. Programme for the Endorsement of Forest Certification (PEFC). Available at: http://pefc.org/news-a-media/general-sfm-news/2015-hungary-becomes-latest-pefc-member

PEFC. 2016c. Green Building's LEED credit promotes sustainable forestry. 7 April 2016. Programme for the Endorsement of Forest Certification (PEFC). Available at: http://pefc.org/news-a-media/general-sfm-news/2071-green-building-s-leed-credit-promotes-sustainable-forestry

Roddy, D. 2016. The science and provincial leadership behind ecosystem-based forestry. 2 February 2016. CBFA News. Available at: http://cbfa-efbc.ca/the-science-and-provincial-leadership-behind-ecosystem-based-forestry/

Rossiyskaya Gazeta. 2013. Federal Law No. 415-FZ of 28 December 2013. Available at: http://rg.ru/2013/12/30/drevesina-dok.html

Rossiyskaya Gazeta. 2015. Federal Law No. 206-FZ of 29 June 2015. Available at: http://rg.ru/2015/07/08/les-dok.html

Sustainable Biomass Partnership 2016a. Sustainable Biomass Partnership timeline. Available at: www.sustainablebiomasspartnership.org/about-us/timeline

Sustainable Biomass Partnership. 2016b. SBP Bulletin. March 2016, Issue 2. Available at: http://us10.campaign-archive1.com/?u=a1d19b4d9b6419dde4023da32&id=00e528a773

Science Committee of the Canadian Boreal Forest Agreement. 2015. Towards a natural range of variation (NRV) strategy for the Canadian Boreal Forest Agreement. Summary report. December 2015. Available at: http://cbfa-efbc.ca/wp-content/uploads/2016/01/NRVSummary_2015FINAL_CompressedEnglish.pdf

SFI 2015. SFI sets a Guinness World Record one tree at a time. 22 September 2015. Sustainable Forestry Initiative (SFI). Available at: www.sfiprogram.org/media-resources/news/sfi-sets-a-guinness-world-record-one-tree-at-a-time-officially-amazing

SFI 2016a. Sustainable Forest Initiative: progress report 2016. Sustainable Forestry Initiative (SFI). Available at: www.sfiprogram.org/files/pdf/2016-sfi-progress-report-summary

SFI 2016b. Program metrics. Personal communication with Rachel Dierolf and James Griffith in May and July 2016. Sustainable Forestry Initiative (SFI).

Taylor, T. 2016. The EU carbon trading market: a decade of experience. 7 April 2016. Available at: http://conversableeconomist.blogspot.com/2016/04/the-eu-carbon-trading-market-decade-of.html?m=1

UNEP DTU. 2016. CDM Pipeline Analysis and Database. Partnership of the UN Environment Programme (UNEP), Danish Ministry of Foreign Affairs, and the Technical University of Denmark (DTU). Available at: http://www.cdmpipeline.org

UNFCCC. 2015. Adoption of the Paris Agreement. Decision 1/CP.21 of the twenty-first session of the Conference of the Parties to United Nations Framework Convention on Climate Change (UNFCCC). December 12, 2015. Available at: http://unfccc.int/meetings/paris_nov_2015/session/9057/php/view/decisions.php

USDA. 2016a. Celebrating five years of the USDA certified biobased product label. 8 March 2016. US Department of Agriculture (USDA). Available at: www.biopreferred.gov/BioPreferred/faces/news/FiveYears.xhtml

USDA. 2016b. BioPreferred catalog. April 2016. US Department of Agriculture (USDA). Available at: www.biopreferred.gov/BioPreferred/faces/catalog/Catalog.xhtml

Van Damme, L., Burkhardt, R., Plante, L. and Saunders, K. 2014. Status report on ecosystem-based management (EBM): policy barriers and opportunities for EBM in Canada. Prepared for the Canadian Boreal Forest Agreement. KBM Resources Group.

World Rainforest Movement. 2016. Destructive biofuels and wood-based biomass out of next Renewable Energy Directive say 120 organisations to EU renewables consultation. Press release 10 February 2016. Available at: http://wrm.org.uy/fr/toutes-les-campagnes/destructive-biofuels-and-wood-based-biomass-out-of-next-renewable-energy-directive-say-120-organisations-to-eu-renewables-consultation

3 PALLETS AND WOODEN PACKAGING

Authors: ***Gunilla Beyer and Marshall White***

HIGHLIGHTS

- Data from the European Federation for Manufacturers of Wooden Packaging indicate that pallet production in 15 European countries increased from 339 million units in 2006 to 401 million units in 2013.
- The pallet and wooden packaging industry consumed more than 20 million m^3 of sawnwood (mostly softwood) in the European subregion in 2015.
- Just fewer than 800 million pallets (both new and rebuilt units) were manufactured in North America in 2011, using about 16.5 million m^3 of new lumber and 12.6 million m^3 of reclaimed lumber.
- Pallet construction is becoming more standardized in Europe, with a migration to the EUR pallet; the 800 x 1,200 mm unit is most prevalent. Pallet sizes and designs are much less standardized in North America, where most pallets in use are custom-designed to suit transport and shipping configurations.
- Pallet pools – whereby companies rent, lease or share the use of pallets – are increasingly used in Europe. Bigger pools are anticipating more pallet repair as the use of pools grow.
- Rental pools are established in North America, and it is anticipated that the movement of food and consumer goods between the US, Canada and Mexico will increasingly be on rental pallets.
- In Europe, the production of pallets and wooden packaging has been shifting to eastern Europe, where costs are lower.
- Pallet prices have been reasonably stable in Europe since 2013; in North America, prices increased between 2013 and 2016 by an average of 3.8% per year.
- Pallets and wooden packaging are well positioned to flourish under policies and laws aimed at a low-carbon economy and sustainability because they have very high rates of reuse, repair and recycling, and they can be used for wood energy or the manufacture of particle board at the end of their useful lives.
- France and the US dominate global barrel exports, with France accounting for about $807 million and the US for $450 million of the value of barrel exports in 2015.

3.1 INTRODUCTION

The classification of wooden packaging includes pallets, pallet collars, box pallets, boxes, crates, cable drums, lightweight packaging for fruit and vegetables, barrels, tailor-made constructions, and dunnage for supporting goods under transportation.

Wooden pallets, crates and packaging cases play important roles in the movement and storage of goods worldwide. Proper design and quality standards ensure that performance is sufficient to protect the goods transported.

Pallets are by far the most common type of wooden packaging. They provide a safe, effective transport and storage platform throughout the handling and distribution process. Although there are standardized versions, pallets are produced in many sizes and configurations to accommodate different handling equipment (generally forklifts), cargoes, space constraints and required longevity (i.e. single or multiple use).

Industrial packaging includes, for example, boxes, crates, box pallets, bins, cages and pallet collars. Pallet collars can function as both pallets and (with a lid and bottom) strong boxes; they are stackable, meaning they can form boxes of varying heights, and many versions are collapsible, meaning they can be stored efficiently when not in use. Pallet collars are used widely, such as for the transport and storage of small parts in the assembly industry. Foldable pallet collars with lids and bottoms can create demountable boxes for simple, cheap return transportation and storage.

Lightweight packaging includes crates, cases, boxes and small drums; it is used mainly for processed or fresh food, beverages, and other consumer goods demanding quality and protection.

Cable drums are used by cable manufacturers in the electrical, electronics and telecommunications sectors.

3.2 EUROPE

3.2.1 CONSUMPTION

Most pallets and wooden packaging in Europe are made of softwood produced in sustainably managed forests; it is estimated that about 4 billion pallets are in circulation in Europe. The average lifespan of a pallet is 5-7 years. The pallet and wooden packaging industry in Europe consumed more than 20 million m^3 of sawnwood in 2015, which was more than 20% of total sawnwood production. When the economy booms, so too does the pallet and wooden packaging industry.

Pallets are manufactured to standards or custom-made. There is a high degree of standardization in Europe based on the modular size of 600 x 400 mm. The major pallet footprints in Europe are 800 x 1,200 mm and 1,000 x 1,200 mm, but half-pallets (600 x 800 mm) and quarter-pallets (400 x 600 mm) are also produced; these tend to be used on full-sized pallets, especially for small shops ordering small quantities. The UK and Benelux export markets, where 1,000 x 1,200 mm pallets were previously the standard, are switching to the EUR pallet of 800 x 1,200 mm.

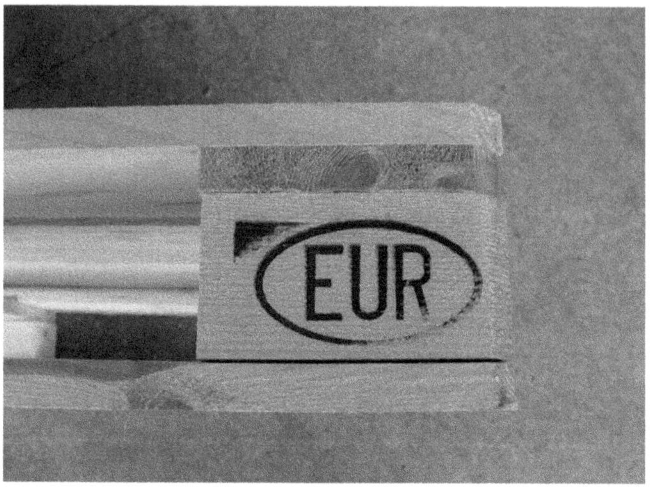

Source: G, Beyer, 2016.

In Europe, four-way-entry block pallets (meaning the forklift can enter from any of the four sides of the pallet) are used most commonly: they now comprise 85% of all pallets, up from 80% in 2006. The prevalence of multi-use pallets increased from 55% in 2006 to 60% in 2013, and the number of pallets repaired increased from 71 million units in 2010 to 129 million units in 2013 (FEFPEB, 2016a).

Pallet pools, whereby companies rent, lease or share the use of pallets, are growing in Europe. The big pools anticipate more pallet repair as the use of pools grows.

3.2.2 Production and capacity change

The production of pallets and wooden packaging is fragmented in Europe, with a large number of small and medium-sized enterprises operating. Production has been shifting to eastern Europe, where costs are lower; pallet manufacture is becoming more automated in western Europe, with the industry incorporating more robotics into production and repair. The sector has consolidated in recent years, with fewer and bigger companies, and a few of the larger groups have started operating internationally.

Production for all the main wooden packaging products increased in Europe from 2012 to 2014, with a total value in 2014 of $11.1 billion (graph 3.2.1).

GRAPH 3.2.1

Production value of flat pallets and pallet collars, EU28, 2012-2014

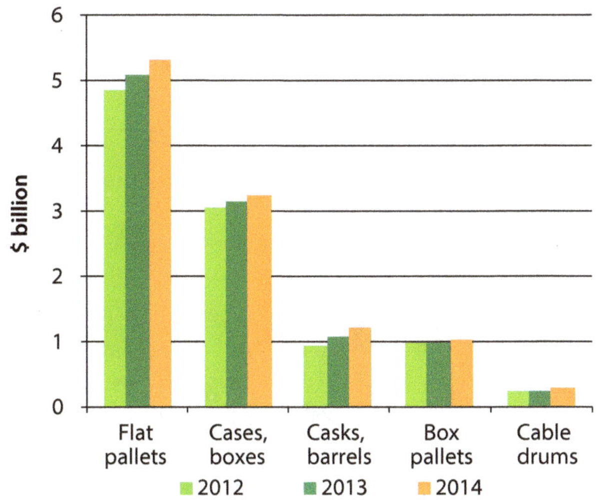

Source: Eurostat, 2016.

3.2.2.1 Flat pallets and pallet collars

France, Poland and Germany are the largest producers (in descending order) of flat pallets and pallet collars in the EU28; production has been increasing at the highest rate in Poland (graph 3.2.2).

GRAPH 3.2.2

Production of flat pallets and pallet collars made of wood, EU28 countries, 2012-2014

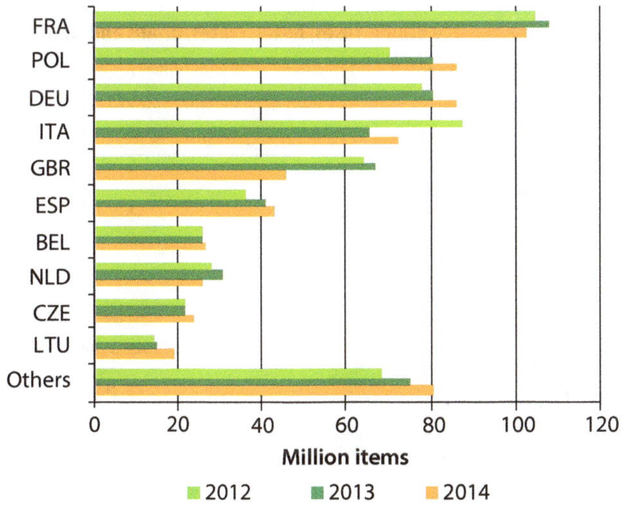

Notes: FRA = France; POL = Poland; DEU = Germany; ITA = Italy; GBR = United Kingdom of Great Britain and Northern Ireland; ESP = Spain; BEL = Belgium; NLD = Netherlands; CZE = Czech Republic; LTU = Lithuania.

Source: Eurostat, 2016.

The European Federation for Manufacturers of Wooden Packaging (FEFPEB) collects statistics on pallet production from its member organizations. Its data, from 15 countries (Austria, Belgium, Denmark, France, Germany, Italy, Lithuania, the Netherlands, Portugal, Spain, Sweden, Switzerland, Turkey, the UK and Ireland) show that the number of manufactured pallets increased from 339 million units in 2006, to 371 million units in 2010 and, to 401 million units in 2013.

Wooden packaging material in the EU (and in the US and many other countries) need to comply with the International Standards For Phytosanitary Measures No. 15 (ISPM 15)[6] standard under the International Plant Protection Convention, which is overseen by FAO. It is an internationally agreed phytosanitary standard for the treatment of all forms of packaging made from solid wood. To comply, wooden packaging must undergo heat treatment or fumigation according to the standard.

FEFPEB reported that 38% of pallets manufactured in 2006 were heat-treated, compared with 60% of those made in 2010 and 50% of those made in 2013. Forty-six percent of pallets manufactured in 2010 were heat-treated and kiln-dried, compared with 41% of those made in 2013. The steep increase in heat treatment after 2006 was due to a pinewood nematode infestation in Portugal, which forced all manufacturers in the country to treat all pallets and wooden packaging, triggering a Europe-wide increase in demand for treated pallets and packaging.

In some countries, especially the Nordic countries, nearly 100% of sawn softwood timber is heat-treated and kiln-dried. In other countries, the pallets are manufactured from pre-cut green timber, and heat treatment takes place after assembly.

Pallet pools of various types are growing in Europe. Many companies are finding benefits in pooling, such as consistent quality, flexibility, the avoidance of capital expenditure, the reduction of costs, more cost certainty, and a reduction in the loss of assets, especially in closed pools.

The most common pallet pool in Europe, the European Pallet Association (EPAL)'s Euro-pallet, originated from wooden pallets used in railway transport.[7] The pallet was standardized in 1961, after which most European industries starting using it, optimizing their trucks, forklifts and high-rack warehouses to do so. Euro-pallet is an open pool involving framework agreements among national associations for pallet exchange: freight is delivered on Euro-pallets, and an equal number of Euro-pallets is provided in return ("pallet for pallet"). It is estimated that nearly 500 million EPAL Euro-pallets are in circulation.

About 73.6 million EPAL Euro-pallets were produced in 2015, up from 67 million in 2012. An estimated 23.9 million units were repaired in 2015, compared with 22.4 million units in 2014 (EPAL, 2016).

In a closed pool, the pallets always remain the property of the pooler, who manages, tracks, recovers, maintains and repairs the pallets. The Commonwealth Handling Equipment Pool (CHEP), the Faber Halbertsma Group Return System (PRS) and La Palette

[6] www.ispm15.com.

[7] EPAL was founded in 1991 as the umbrella association of licensed producers and repairers of EPAL-pallets and box pallets.

Source: Beyer, G. 2016.

Rouge (LPR) are major closed pools; together they account for roughly 200 million pallets.

CHEP evolved from the Allied Materials Handling Standing Committee, an organization developed by the Australian Government for the efficient handling of defence supplies in the Second World War. The largest CHEP pallet-pool countries in the EU are (in descending order) the UK, Spain, France, Germany and Italy (CHEP, 2016).

Pooling Partners (in the Faber Halbertsma Group) is a pooling-services provider, and it also manufactures pallets and boxes. It operates three pooling networks: International Pallet Pool Logipal B.V. (IPP Logipal); PAKi Logistics; and the PRS Return System. The PRS Return System was established in 1997 to rent pallets to the chemical industry when it adopted pallet standardization. It is estimated that Pooling Partners moves more than 75 million pallets per year (IPP Logipal, 2016).

The LPR was established in 1989 in France as a pallet-pooling specialist, and it has expanded progressively across Europe, with a focus on the consumer goods supply chain. LPR reported 73 million pallet movements in 2015 (LPR, 2016).

Another major pooling provider is the Palettes Gestion Services (PGS) Group; this is mainly a business-to-business pool, with more than 25 million pallets (PGS Group, 2016).

3.2.2.2 Box pallets and load boards of wood (excluding flat pallets)

The production of box pallets and load boards was fairly stable in the EU28 from 2012 to 2014, with about 112 millions items produced in 2014. Among EU28 countries, Italy dominates production (graph 3.2.3).

GRAPH 3.2.3

EU28 production of box pallets and load boards of wood, 2012-2014

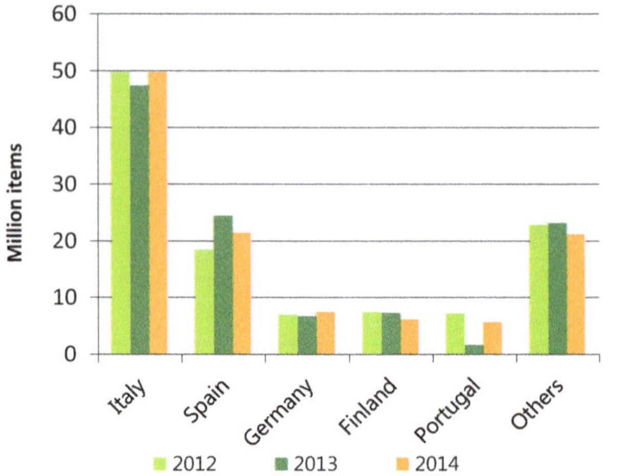

Source: Eurostat, 2016.

3.2.2.3 Cases, boxes, crates, drums and similar wooden packaging

There was an increase in the production of cases, boxes, crates, drums and similar wooden packaging in Europe from 2012 to 2014. Italy was the leading producer, followed by Spain and France (graph 3.2.4).

The production of cable drums increased by 50% in 2014 compared with 2013, with Hungary increasing production significantly.

GRAPH 3.2.4

EU28 production of cable drums, cases, boxes, crates, drums and similar packages of wood, 2012-2014

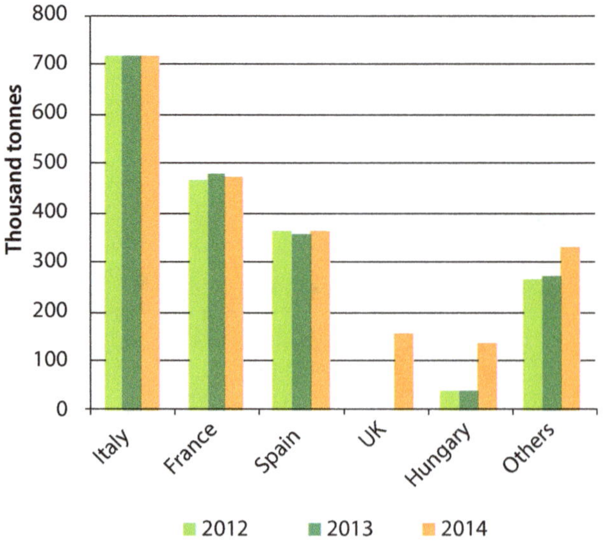

Source: Eurostat, 2016.

3.2.2.4 Lightweight packaging

Lightweight packaging, used mainly for fruit and vegetables, is reported separately in the FEFPEB data. Table 3.2.1 and graph 3.2.5 show the number of units manufactured in 2010 and 2013 for those countries that provided data to FEFPEB; Spain was the largest manufacturer, followed by France. There was a slight decrease in production in France between 2010 and 2013 and an 18% increase in Spain. Roughly 80% of the lightweight packaging produced in Spain in 2013, and all the production in Portugal, were reported to be heat-treated in conformity with ISPM 15.

TABLE 3.2.1

Wood use for lightweight packaging reported to FEFPEB, 2010 and 2013

	2010	*2013*
Million units	1,095	1,171
m³	1,232,891	1,243,000

Source: FEFPEB, 2016a.

GRAPH 3.2.5

Lightweight wooden packaging production, top five EU28 countries, 2010 and 2013

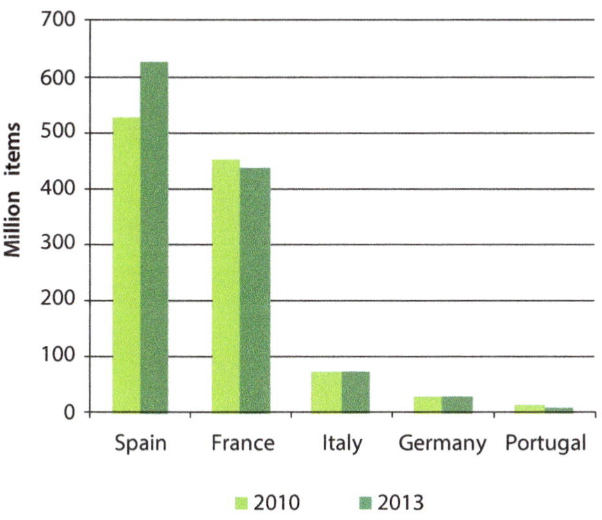

Source: FEFPEB, 2016a.

3.2.2.5 Barrels and cooperage products, including staves

Casks, barrels, vats, tubs and other cooperage products and parts, including staves, tend to have high values, making them quite cost-effective for long-distance trade. Wine barrels, for example, are made mostly of oak.

The main producers of barrels and cooperage products in the EU28 are (in descending order) France, Spain, Italy and Hungary. Production in France, the predominant producer, increased by 170% between 2012 and 2014; over the same period, the increase for the EU28 as a whole was about 125% (graph 3.2.6) France is the world's largest exporter of barrels, exporting $807 million worth in 2015.

GRAPH 3.2.6

Production of wooden casks, barrels, vats, tubs and other cooperage products and parts, including staves, EU28, 2012-2014

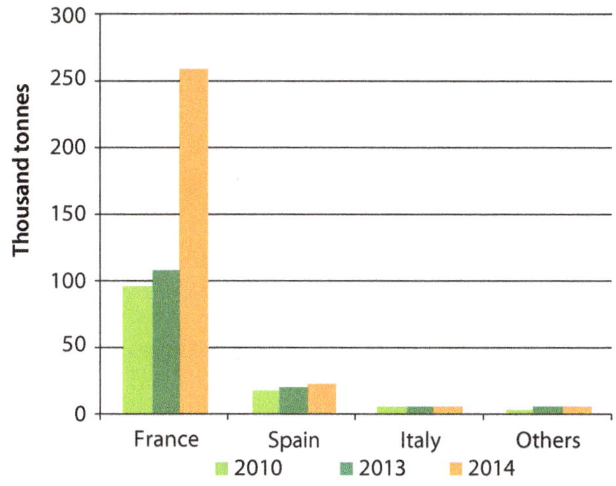

Source: Eurostat, 2016.

3.2.3 Prices

Pallet prices have been fairly stable in recent years, even though prices of raw materials have fluctuated (graph 3.2.7). There is price pressure on pallets, and overcapacity has been reported.

GRAPH 3.2.7

FEFPEB pallet timber price index, 2009-2015

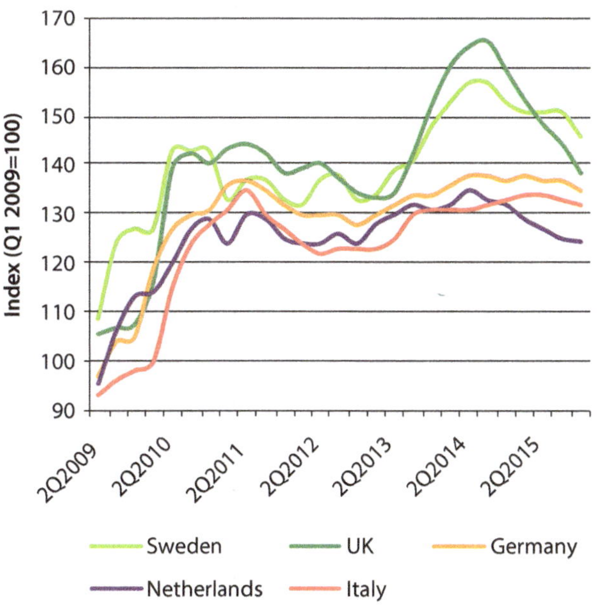

Notes: This figure has been produced from the national timber price indices of Germany, Italy, the Netherlands, Sweden and the UK, using comparable pallet wood sizes and a common base year (2009) to make them comparable. The FEFPEB pallet timber price index indicates price trends for the wood used for pallets in Europe.
Source: Eurostat, 2016.

3.2.4 Trade

Pallets and wooden packaging are mostly shipped with goods, but there is also some trade of empty pallets and wooden packaging, mainly within Europe.

3.2.4.1 Flat pallets and pallet collars

This sector is developing strongly in Poland and other eastern EU countries (graph 3.2.8). Many western European countries find it hard to compete with eastern European prices; imports are increasing in western Europe and production is moving to low-cost countries. Of the EU28 countries, Germany is the largest importer of pallets and pallet collars (graph 3.2.9).

GRAPH 3.2.8

Exports of flat pallets and collars of wood, EU28, 2012-2014

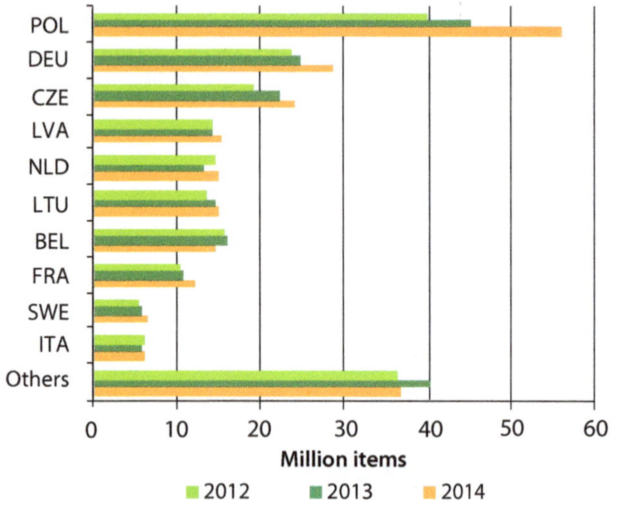

Notes: Pol = Poland; DEU = Germany; CZE = Czech Republic; LVA = Latvia; NLD = Netherlands; LTU = Lithuania; BEL = Belgium; FRA = France; SWE = Sweden; ITA = Italy.
Source: Eurostat, 2016.

GRAPH 3.2.9

Imports of flat pallets and collars of wood, EU28, 2012-2014

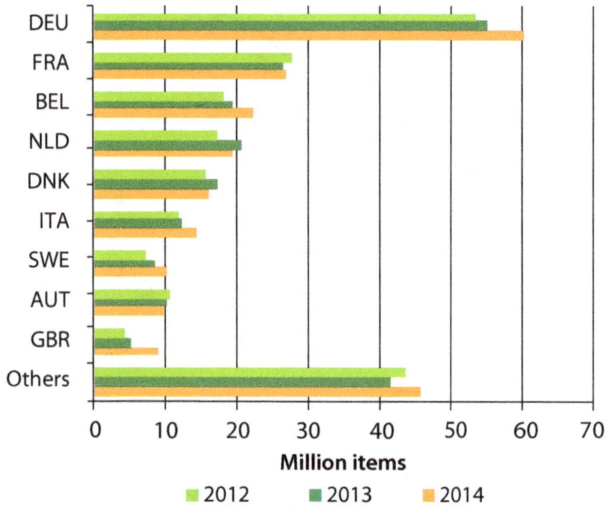

Notes: DEU = Germany; FRA = France; BEL = Belgium; NLD = Netherlands; DNK = Denmark; ITA = Italy; SWE = Sweden; AUT = Austria; GBR = United Kingdom of Great Britain and Northern Ireland.
Source: Eurostat, 2016.

The EU28 countries trade pallets and wooden packaging (excluding casks and barrels) more strongly with each other than extraregionally. Barrels and casks dominate exports, making up about half the value of all imports in the pallets and wooden packaging category. Exports in 2014 for all other categories except pallets (which increased by 12%) fell in 2014 compared with 2013, and imports increased substantially (table 3.2.2).

TABLE 3.2.2

EU28 trade of pallets and wooden packaging, 2012-2014
($ million)

	2012	*2013*	*2014*	*Change (%) 2013-2014*
		Exports		
Flat pallets	129	137	155	13.1
Box pallets	54	56	50	-10.7
Casks, barrels	351	414	399	-3.6
Cases, boxes	62	77	76	-1.3
Cable drums	19	20	17	-15.0
		Imports		
Flat pallets	72	79	107	35.4
Box pallets	22	24	27	12.5
Casks, barrels	112	136	178	30.9
Cases, boxes	33	33	39	18.2
Cable drums	5	4	6	50.0

Note: Figures rounded to the nearest $ million. Includes EU28 intraregional trade.
Source: Eurostat, 2016.

3.2.5 European policy and regulatory influences

3.2.5.1 Circular economy

In December 2015, the European Commission launched a communication on the "circular economy", which has now been sent out to EU member states for consultation. The EU's action plan for a transition to a more circular economy aims to develop a sustainable, low-carbon, resource-efficient and competitive economy (the targets are under discussion and therefore not yet decided). The pallet and wooden packaging sector is well placed to thrive in such an economy because its products have a very high rate of reuse, repair and recycling and can be used to generate wood energy or in the manufacture of particle board at the end of their useful lives.

Legislative proposals on waste, adopted along with the action plan for a transition to a more circular economy, include long-term targets to reduce landfilling and increase the preparation of materials for reuse and recycling, as well as higher recycling targets for packaging materials. The revised waste proposals will also address key issues relating to the calculation of recycling rates, but it is unclear how the number of pallet trips are to be verified and calculated.

The targets laid down in the earlier (1994) Directive 94/62/EC of the European Parliament and of the Council for the recovery and recycling of packaging and packaging waste is being amended to increase reuse and recycling in order to better reflect the EU's packaging waste ambition in moving towards a circular economy. The proposal is as follows: no later than 31 December 2025, a minimum of 65% (increased to 75% at the end of 2030) by weight of all packaging waste will be prepared for reuse and recycling, with a minimum target of 60% for wood (European Commission, 2016).

3.2.5.2 Other policy issues with impacts on pallets and wood packaging

Many ongoing and potential policy issues are affecting the pallet and wooden packaging sector. Environmental product declarations are now often requested, and both users and policymakers are concerned about the carbon footprints of products. The wooden pallet and packaging sector is also well-placed in this area, comparing favourably with competing materials used for pallets and packaging.

Purchasers of wood for the manufacture of pallets and wooden packaging in the EU must comply with the EU Timber Regulation. Producers must exercise due diligence and keep records on wood sources.

The use of wood as packaging for food has faced regulatory and perceptional hurdles due to the possibility of splinters, wood's porosity, and the view that wood is difficult to clean and sanitize. Two recent studies in France and Spain, however, have demonstrated the superior antimicrobial properties of wood species such as pine and poplar compared with smooth synthetic materials, including plastics (FEFPEB, 2016b).

3.3 COMMONWEALTH OF INDEPENDENT STATES, WITH A FOCUS ON THE RUSSIAN FEDERATION

Demand for wooden pallets grew by more than 82% in the Russian Federation between 2010 and 2014, from 9.8 million to 17.8 million units. The biggest growth was in 2011 and 2012, the result of increases in cargo movements. Growth in demand slowed in 2013, to 5%, and the first decline in demand for five years was recorded in 2014 (RBC, 2016). In line with the economic downturn in the Russian Federation, demand for wooden pallets was forecast to decline further in 2015, by 2.4%, but a rebound is expected in 2016-2019 (RBC, 2016).

The vast majority of wooden pallets produced in the Russian Federation are used domestically. Domestic sales accounted for about 95% of total production by volume in 2010-2014, with exports accounting for the remainder. Wooden pallet producers have focused on developing the internal market: domestic sales grew by more than 86% from 2010 to 2014, while exports grew by only 31.5% (RBC, 2016).

The largest international buyers of Russian wooden pallets in 2014 were Belarus (295,000 units), Germany (262,000 units) and Lithuania (78,000 units) (RBC, 2016).

3.4 NORTH AMERICA

Eighty percent of consumer and industrial products moving along North American domestic supply chains are palletized. Pallet sizes and designs are less standardized in North America than in other industrialized regions, with most of those in use custom-designed for specific product shipments through designated supply chains. Unlike in other regions, too, a large number of businesses specialize in collecting, repairing and recycling pallets, whether or not they are reusable. Pallets are specified in commerce in US customary units.

Source: G. Beyer, 2016.

3.4.1 Consumption

3.4.1.1 Pallets

In 2011, 742 million wooden pallets were made in the US, and between 50 million and 55 million were made in Canada. Of those made in the US, 56% were manufactured new and 44% were used, repaired or remanufactured.[8] Wooden pallets represent 90-94% of the manufactured pallets in North America.[9] It is estimated that about 2 billion pallets are in use in domestic supply chains, of which about 100 million are rental pallets provided by CHEP and PECO.[10] The manufacture of pallets in the US consumed 16.5 million m^3 of new lumber and 12.6 million m^3 of reclaimed lumber and parts in 2011. Sixty-three percent of the wood used was softwoods such as pine, hemlock, spruce, fir and Douglas fir, and the other 37% comprised the hardwoods oak, maple, birch, yellow poplar and a mix of other species, such as red alder and aspen. An estimated 13.3% of the wood was certified (Bush and Araman, 2015).

8 A remanufactured pallet is a pallet assembled with used parts.

9 An estimated 4-6% are plastic pallets and 5% or less are paper and metal pallets.

10 PECO Pallets (www.pecopallet.com) is a provider of pallet-pooling services and logistics in North America.

Graph 3.4.1, which shows pallet sizes and design by market share in the US, reflects the relatively low level of pallet size and design standardization in North America. Although the 48 x 40 inch and 40 x 48 inch pallet sizes have the same plan size, they have different designs. More than half the pallets in use are in dimensions representing less than 1% of pallets sold by manufacturers or rebuilders (Bush and Araman, 2015).

GRAPH 3.4.1

Pallet sizes manufactured in the US, by industry sector, 2011

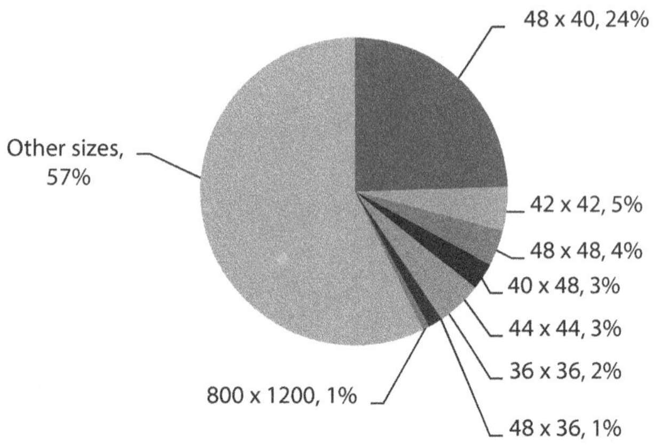

Note: Sizes in inches except 800 x 1200, which is in millimetres.
Source: Bush and Araman, 2015.

Some 2,270 firms in the US and 489 firms in Canada were producing wood containers and pallets in 2015.[11] The total estimated value of the product of these firms in 2015 was $7.7 billion in the US and $579 million in Canada.

The food and related industries are the largest users of pallets in North America, followed by the chemical and pharmaceutical industries, and then agriculture (IBIS World, 2015) (graph (3.4.2).

GRAPH 3.4.2

Pallet use, by industry

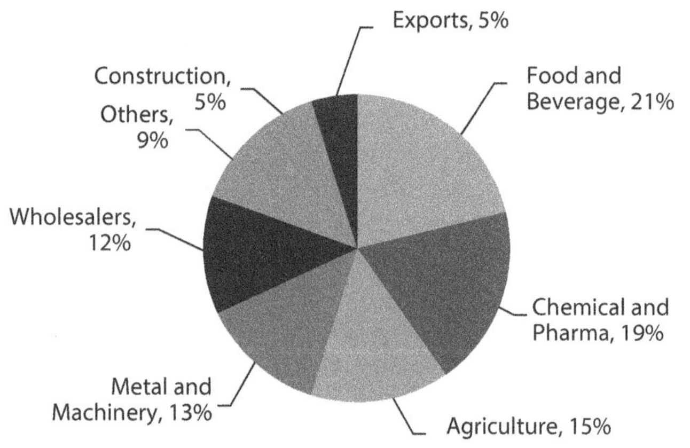

Source: IBIS World, 2015.

3.4.1.2 Barrels and cooperage products, including staves

The barrel industry has been booming in the US, driven largely by a 50% increase in the production of bourbon whiskey between 2010 and 2013 (Mickle, 2015). The US exported about 49,000 tonnes of barrels in 2015 (more than double the quantity in 2011), with a value of about $450 million.

A 2015 survey of the cooperage industry conducted by the Associated Cooperage Industries of America (ACIA) showed that the overwhelming number of responding companies expected demand to stay strong in 2016. The industry supplies new barrels for the production of wine and bourbon, with many used barrels having second lives in the production of scotch whisky (ACIA, 2015).

Source: Esperanza33, 2016.

Wine and spirits are often aged in barrels for extended periods, creating a significant lag between barrel demand and demand for wine and spirits. In 2014, more barrels of whisky were aging in Kentucky (population 4.41 million) than there were residents in the state (Marchetti, 2015).

3.4.2 Production and capacity change

Table 3.4.1 shows that there was consistent growth in the reuse of recovered, repaired and remanufactured pallets in the US between 1995 and 2011 (Bush and Araman, 2015). The ratio of new-pallet to reclaimed-pallet production is expected to stabilize in the future as the condition of used pallets declines and new replacements are needed.

TABLE 3.4.1

Estimated production of new and recovered/repaired/remanufactured pallets in the US, 1995-2011

(millions of units)

	1995	1999	2006	2011
New	411	429	441	416
Recovered/ repaired/ remanufactured	143	223	321	326

Source: Bush and Araman, 2015.

[11] This refers to the North American Industry Classification System, NAICS 321920 Wood Container and Pallets; see census.gov.

Revenue in the US pallet-manufacture sector contracted dramatically (by 20.6%) in 2009, corresponding with the economic recession. Annual revenue growth was consistently between 3% and 4% from 2010 to 2013, after which it stabilized (IBIS World, 2015) (graph 3.4.3).

GRAPH 3.4.3

Revenue growth in the US pallet sector, 2008-2015

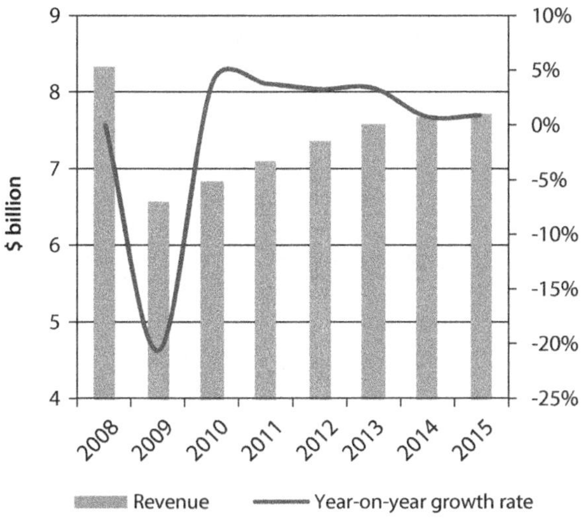

Source: IBIS World, 2015.

3.4.3 Prices

North American wooden pallet prices increased by an average of 3.8% per year from 2013 to 2016. Prices for hardwood pallet lumber increased significantly for much of the period and have stabilized recently, while prices for softwood pallet lumber have declined somewhat (Pallet Profile Weekly, 2016).

3.4.4 Trade

3.4.4.1 Imports

North America is involved in very little international sale and movement of empty pallets because of the high cost of transportation and the low marginal value of pallets. Pallets coming into North America under product are recovered and sold domestically. The importation of pallets under product has increased in recent years by an annual average of 1.8%, and its estimated value in 2015 was $489.9 million; nevertheless, imported pallets meet only about 6.5% of domestic demand in North America. There is a growing cottage industry that buys 800 x 1,200 mm Euro-pallets and resells them to manufacturers shipping to Europe. Most product arriving in North America is floor-loaded into containers and not palletized until after arrival. Of the pallets imported into North America, 47% comes from France, 22% comes from China, 17.4% comes from Canada and about 2% comes from Mexico (IBIS World, 2015). It is expected that, as the US dollar appreciates, there will be a small increase in pallet imports. The CHEP and PECO pallet rental pools are both establishing in Canada and Mexico, and an increasing proportion of the food and consumer goods moved between the US and these countries will be on rental pallets in the future.

3.4.4.2 Exports

The value of exported pallets in 2015 was estimated at $372.7 million, with the UK and Canada accounting for 53.4% of this trade between them. Annualized growth of exports of 5.8% was predicted but will likely be lower than that due to the increased value of the US dollar, slower-than-expected growth in the China market, and competition from pallets made of other materials (IBIS World, 2015). The ISPM 15 regulation, which requires the treatment of solid wooden packaging materials, including pallets, and is applied in international supply chains, has caused an increase in the cost of wood pallets. Exporters are therefore increasingly using pallets made of non-regulated materials, such as corrugated paperboard, wood-based composites, and plastics.

3.4.5 Extraregional influences affecting North America

Pallet-pooling between Asia and North America is growing. It started in the Korean automotive sector with the delivery of parts from the Republic of Korea to Hyundai and Kia assembly plants in the US; most of the pallets used for this purpose are returned empty to the Republic of Korea. Pallet-pooling could be expanded to the electronics industry in intercontinental shipments of parts, subassemblies and final assemblies, enabling the two-way movement of pallets under product. A barrier to this, however, is the different pallet sizes in use in Asia and North America. The EU, China and the Republic of Korea use 1,200 x 1,000 mm as the standard pallet size, and it is also the most common size in Central and South America. The 1,219 x 1,016 mm dimensions used in North America are close to this size, and the true 1,200 x 1,000 mm pallet might fit many domestic North American supply chains. If this is true, the US could assimilate and reuse many pallets coming in under product from other regions without costly re-palletization or return to countries of origin. The result would be more reusable pallets entering the US and a subsequent reduction in demand for domestically manufactured pallets.

The expansion of the Panama Canal is expected to result in additional warehousing capacity in the eastern US. The larger 12,000 twenty-foot equivalent units[12] vessels arriving in eastern ports will cause a regional shift in North American pallet demand, with demand increasing in the east and intermodal movements from the west to the east declining.

[12] An inexact unit of cargo capacity used for container ships and terminals, based on the volume of a 20-foot-long (6.1 m) intermodal container.

Source: G. Beyer, 2016

3.4.6 North American policy and regulatory influences

The US's 2008 Food Safety Modernization Act expanded the authority and responsibility of the Food and Drug Administration (FDA) beyond food manufacturing to include the entire food product supply chain. The sanitation of pallets is a growing focus of the FDA, which has increased pallet inspections for cleanliness and damage. The impact on the market of such greater scrutiny will be the increased use of dry pallets and of treatments to control moulds and insect infestations. Pallet washers are now being used to clean reusable wood pallets. The increased use of plastic and metal pallets, which are perceived to be more sanitary, is having an impact on the wooden pallet market. Wooden pallets still dominate the North American market, but the use of plastic pallets is increasing at a greater rate.

The incidence of fire at pallet-manufacturing facilities has increased. In response to this, proposals have been made to regulate the way in which pallets are stored at these locations after assembly. State fire marshals are requiring greater lane space between pallet stacks and between pallet stacks and buildings, which inevitably increases the cost of pallet manufacturing.

Several years ago, the federal government considered legislating to favour the use of plastic pallets by government agencies. Recently, the state of Oregon considered a bill that would give preference to paper pallets for state government use. The justification in both instances was that the use of these alternatives reduced the impact of pallet use on the environment. Both initiatives failed, however.

A bilateral treaty between the US and Canada permits pallet movement between the two countries without compliance with ISPM 15. The expiration of this agreement is being planned, but no timetable has been set. When it does expire, significantly more North American pallets will have to be heat-treated or fumigated. The percentage of new pallets that are heat-treated in the US varies by region, from a low of 30% to a high of 57%, with the highest percentages in the west and east (i.e. the exporting areas) and the lowest in the midwest (Bush and Araman, 2015).

3.5 INNOVATION

3.5.1 Smaller pallets

There is a trend towards the use of smaller pallets for display-ready packaged product to be placed directly on store floors. This includes the so-called half pallet (24 x 40 inches and 600 x 800 mm) and quarter pallet (24 x 20 inches and 400 x 600 mm). Manufacturers palletize their product on these small pallets, which then pass through the supply chain to retail.

3.5.2 Smarter pallets

Because robotic systems do not adjust well to variation in their operating environments, "smarter" pallets are required as supply chains become more automated. Smarter pallets are pallets that are stiffer and do not sag in storage racks; are more uniform in geometry and dimension; and have flat surfaces for better interfacing with packaging and equipment. The detailed description of pallets to be used in automated materials handling systems can be found in ANSI MH1 2016: "Pallets, slip sheets and other bases for unit loads".

3.5.3 Systems-based design of global supply chains

To significantly improve the operational efficiency of global supply chains, supply-chain owners and operators must integrate the design of pallets, packaging and unit-load handling equipment. Today, these three components of the supply chain are designed by three different design communities that do not interact, meaning that supply chains operate with significant avoidable costs. What is needed is a fundamental shift from the "component-by-component" design process to a true "systems" design process that considers how the pallet, packaging, and shipping, storage and handling systems interact mechanically. The pallet is the key because it is the interface between the other two components. The pallet can be used to significantly reduce supply-chain operating costs, improve supply-chain operating safety, and increase supply-chain operational sustainability (White, 2016).

3.6 REFERENCES

ACIA (Associated Cooperage Industries of America, Inc.). 2015. 2015 ACIA survey. Louisville, Kentucky, USA.

Bush, R.J. and Araman, P.A. 2015. Trends in the use of wood products for distribution packaging. USDA Forest Service, Southern Research Station, Blacksburg, Virginia, USA.

CHEP (Commonwealth Handling Equipment Pool). 2016. Available at: www.chep.com

EPAL (European Pallet Association e.V.). 2016. Available at: www.epal-pallets.org

European Commission. 2016. COM(2015)614/2: Available at: http://ec.europa.eu/transparency/regdoc/?fuseaction=list&coteId=1&year=2015&number=614&version=ALL&language=en

Eurostat. 2016. Available at: http://ec.europa.eu/eurostat/web/prodcom/data/database

FEFPEB (European Federation for Manufacturers of Wooden Packaging). 2016a. European Federation for Manufacturers of Wooden Packaging. Available at: www.fefpeb.eu

FEFPEB (European Federation for Manufacturers of Wooden Packaging). 2016b. Available at: http://www.fefpeb.eu/news/new-evidence-confirms-hygiene-of-wood-in-contact-with-fruits-fish-and-cheese

LPR (La Palette Rouge). 2016. Available at: www.lpr.eu

Marchetti, N. 2015. Bourbon barrels easily outnumber Kentucky humans, horses. The Whiskey Wash. Available at: http://thewhiskeywash.com/bourbon/bourbon-barrels-easily-outnumber-kentucky-humans-horses

Pallet Profile Weekly. 2016. Industrial reporting. Richmond Virginia, USA. Available at: palletprofile.org

IBISworld. 2015. Industry, and business research reports and information. 32192 Wood pallets and skids production in the US. Available at: IBISworld.com

IPP (International Pallet Pool) Logipal B.V. 2016. Available at: www.ipplogipal.com

Mickle, T. 2015. Bourbon feels the burn of a barrel shortage. Wall Street Journal. Available at: http://www.wsj.com/articles/bourbon-makers-feel-the-burn-of-a-barrel-shortage-1431371621

PGS (Palettes Gestion Services) Group. 2016. Available at: www.pgs-international.com

RBC (Research Marketing Agency). 2016. Analysis of the market of wooden pallets in Russia in 2010-2014, forecast for 2015-2019 years. Available at: http://marketing.rbc.ru/research/562949982663928.shtml

White M.S. 2016. Systems based design of global supply chains. Proceedings of TransPack 2016. Orlando Florida, International Safe Transit Association, East Lansing, Michigan, USAS.

4 WOOD RAW MATERIALS

Authors: ***Håkan Ekström***

HIGHLIGHTS

- The consumption of roundwood in the UNECE region, comprising logs for industrial uses and fuel, increased by 1.2% in 2015 to reach the highest level in almost ten years.
- Removals of industrial roundwood were up by 2% in Europe in 2015, by 1.1% in the CIS subregion and by 0.1% in North America.
- Timber harvests in the CIS subregion increased by almost 10% in the five years to 2015.
- The UNECE region continues to be a net exporter of logs; shipments of softwood logs from North America and the Russian Federation to China and the Republic of Korea are among the world's largest log trade flows.
- Log consumption in Europe reached its highest level since 2007 in 2015, with the largest increases in the last five years occurring in Finland, Poland, Sweden and Turkey.
- The log market in the Baltic Sea area is one of the most active in the world, with the major trade flows going from the Baltic States to Finland and Sweden.
- Portugal and Turkey are the only European countries importing wood chips from outside the subregion.
- Timber harvests increased in the Russian Federation in 2015; all the increase was in the eastern provinces.
- Log demand by the sawmill sector in North America increased in the five years to 2015, but log consumption by pulp mills declined.
- Softwood sawlog prices mostly declined in 2014 and 2015 in all the UNECE subregions because of a stronger US dollar and reduced global demand for lumber.
- Wood fibre costs for pulp mills in Europe, North America and the Russian Federation trended downward for most of 2015 and early 2016, reaching levels not seen for eight years.

4.1 INTRODUCTION

The total consumption of roundwood – comprising logs for industrial uses and fuel – in the UNECE region was estimated at 1,280 million m³ in 2015, an increase of 1.2% from 2014 and the third consecutive year of growth. Total log use reached its highest level in almost ten years in each of the three UNECE subregions in 2015 due to a steady increase in demand for both industrial roundwood and fuel (graphs 4.1.1 and 4.1.2).

Removals of industrial roundwood in the UNECE region were up by 0.9% in 2015, with the biggest increase in Europe (+2%) and practically no change in North America. Although log production in the CIS subregion was only 1.1% higher in 2015 than in 2014, the longer-term trend is more impressive, with 2015 removals almost 10% above those in 2011. Almost all the increase in the timber harvest in the UNECE region in 2015 was of coniferous logs, with removals of non-coniferous logs steady.

GRAPH 4.1.1

Apparent consumption of softwood industrial roundwood in the UNECE region, by subregion, 2011-2016

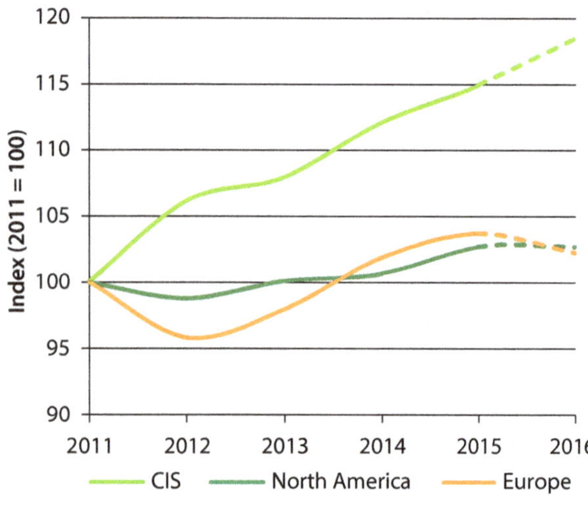

Note: f = 2015 Committee on Forests and the Forest Industry forecast.
Source: UNECE/FAO, 2016.

Of the total roundwood removals in the UNECE region in 2015, approximately 16% (204 million m³) was used for fuel. This volume was consumed predominantly in Europe, which accounted for almost 58% of total woodfuel consumption in the UNECE region. Data for roundwood removals for fuel are unreliable because few countries have consistent methods of collecting relevant data for this increasingly important end use; nevertheless, it is clear that a fairly large share of forest removals are used for energy. This chapter focuses mainly on the production, consumption, trade and prices of industrial roundwood rather than total roundwood (which would include woodfuel); Chapter 9 examines trends for wood raw materials in the wood energy sector.

GRAPH 4.1.2

Apparent consumption of hardwood industrial roundwood in the UNECE region, by subregion, 2011-2016

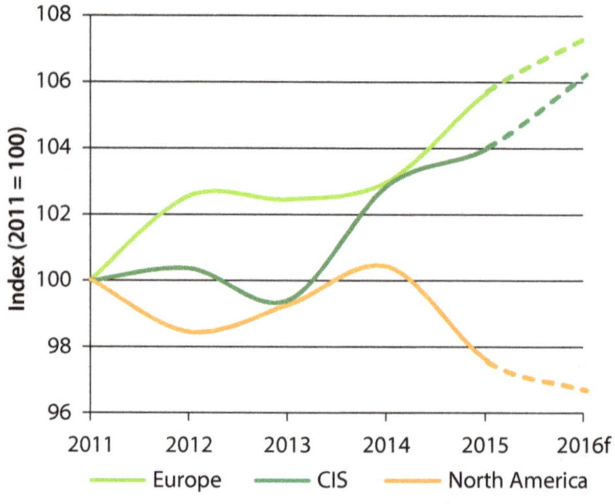

Note: f = 2015 Committee on Forests and the Forest Industry forecast.
Source: UNECE/FAO, 2016.

The global trade of softwood logs fell by almost 10% in 2015, to just over 76 million m³ (Wood Resources International, 2016b), due predominantly to lower demand for wood raw materials in the major Asian markets. Demand for imported logs picked up in early 2016, however, with higher volumes shipped to both China and Japan in the first five months of 2016 compared with the same period in 2015.

Source: UNECE/FAO, 2016.

The UNECE region continues to be a net exporter of logs, with globally significant trade flows of softwood logs from North America and the Russian Federation to China and the Republic of Korea. Of the top five trade flows of softwood logs worldwide, shipments to China from New Zealand, the Russian Federation and the US were all lower in 2015 compared with 2014 (graph 4.1.3).

GRAPH 4.1.3

Top five global trade flows of softwood roundwood, 2011-2015

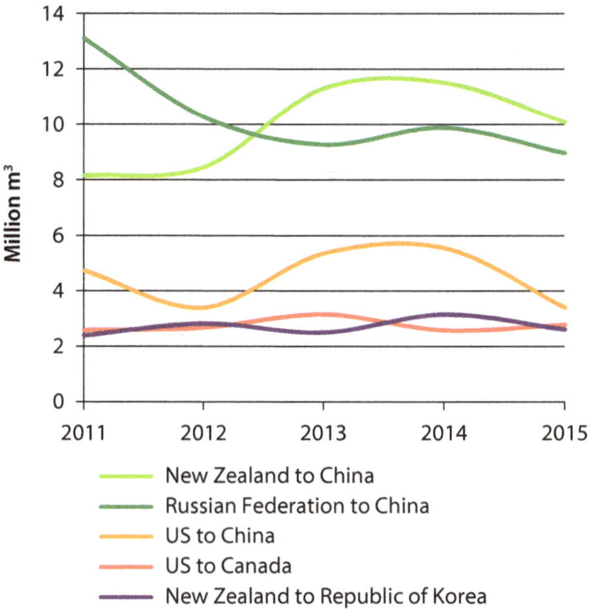

Source: Wood Resources International, 2016a.

TABLE 4.2.1

Industrial roundwood balance, Europe, 2014-2016

(thousand m^3)

	2014	2015	2016f	Change (%) 2014-2015
Removals	381,617	389,205	387,478	2.0
Imports	56,588	55,463	54,927	-2.0
Exports	44,448	43,025	44,266	-3.2
Apparent consumption	393,757	401,642	398,139	2.0

Note: f = 2015 Committee on Forests and the Forest Industry forecast.
Source: UNECE/FAO, 2016.

Source: UNECE/FAO, 2016.

4.2 EUROPE

4.2.1 Industrial roundwood markets

Removals of industrial roundwood in Europe increased for the third consecutive year in 2015, reaching almost 390 million m^3 (Table 4.2.1). The harvest of coniferous logs (about 75% of the total) was up by 1.6% over 2014, while the harvest of non-coniferous logs (about 25% of the total) increased by 3.4%. Although not all European countries have reported their 2015 timber harvest volumes, a few stand out among those who have. The five largest log-producing countries in Europe in 2015 were, in descending order by volume, Sweden, Finland, Germany, Poland and Turkey, all of which increased production compared with 2014. With the exception of Sweden, most of the increase in log removals was of coniferous logs.

European demand for logs in 2015 followed the same trend as for log production, with log consumption reaching its highest level since 2007. The biggest increases in log use by domestic forest industries in the five years to 2015 were in Finland (+5.9 million m^3), Turkey (+3 million m^3), Poland (+2.7 million m^3) and Sweden (+2.6 million m^3). On the other hand, log consumption decreased in the five-year period in Norway (-2.4 million m^3), Italy (-1.1 million m^3) and France (-0.9 million m^3). The main reasons for the changes in log consumption have been higher lumber production (northern Europe and Poland), a rise in the production of composite board (Turkey), reduced pulp production (Norway), lower lumber production (France), a decline in composite board manufacturing (Italy), and, in general, fewer roundwood imports.

4.2.2 Trade of roundwood and wood chips

Europe's log trade declined in 2015 despite increased demand for wood raw materials. The higher demand for logs was met instead by a rise in domestic harvesting and a reduction in log exports. Imported logs met 13.8% of total roundwood demand in 2015, down from 15.1% in 2013; net imports fell from 14.5 million m^3 to 12.5 million m^3 over the same period. Some of the biggest changes in log trade flows the past few years have been in the Nordic countries and in central Europe.

The Baltic Sea area is one of the world's most active log markets, accounting for more than 20% of the global softwood log trade in 2015 and almost 29% of the global trade of temperate hardwood logs. Finland and Sweden are the major importing countries, but forest companies in Germany and Poland also imported substantial log volumes in the five years to 2015.

Log import volumes to the Nordic countries reached a six-year high in 2014, at 14.3 million m^3, but shipments to those countries fell by more than 10% in 2015, due mainly to lower demand for softwood pulplogs. On the other hand, the trade of softwood sawlogs to Finland and Sweden reached its highest level since 2007 in 2015, when 1.6 million m^3 was imported, primarily by sawmills in Sweden.

A decade ago, the Russian Federation exported about 7.5 million m^3 of softwood logs to the Nordic countries, but shipments plunged after the country's introduction of log export duties, and the Russian Federation has exported just over 1 million m^3 to the Nordic countries for each of the last three years. As Russian log exports have fallen, log exporters in the Baltic States have stepped in, and that subregion is now the major supplier of logs to sawmills and pulp mills in the Nordic countries and Germany. Shipments of softwood logs from the Baltic countries have declined steadily, however, from 3.1 million m^3 in 2011 to about 1.3 million m^3 in 2015. Exports of hardwood logs from the Baltic States also fell (by almost 30%) in the five years to 2015, but the Russian Federation has become a more aggressive player as the weak rouble has made its logs more competitive.

Germany and Austria are respectively the world's second-largest and third-largest importers of softwood logs; Germany in particular has increased its imports substantially in the last five years. In 2008, Germany was a net log exporter by about 1.6 million m^3, but the flow of softwood logs has turned around since and the country was a net log importer (by 5.4 million m^3) in 2015. The major log-supplying countries in 2015 and early 2016 were (in descending order, by volume) the Czech Republic, Poland, Norway and Estonia.

Portugal and Turkey are the only European countries that imported wood chips from outside Europe in 2015; end uses are hardwood pulp in Portugal and MDF and particle board in Turkey.

Pulp mills in Portugal have been importing fairly large volumes of eucalyptus chips from Latin America since 2008, when the first shipments arrived from Chile and Uruguay. Shipments peaked in 2011, at 526,000 oven-dry tonnes, and then declined sharply to 212,000 oven-dry tonnes in 2012. Shipments of eucalyptus chip have increased annually in the last three years, reaching 451,000 oven-dry tonnes in 2015. Uruguay supplied about 72% of that volume, and the remainder originated in Brazil.

4.2.3 Consumption of wood fibre by the pulp industry

Pulp production in Europe in 2015 was practically unchanged from 2014, which was the lowest level since 2009. The pulp industry has consumed less wood fibre in the last few years; it used almost 145 million m^3 of roundwood and wood chips in 2015, which was down by 1.5% from 2014 and 3% below the ten-year average.

Just over 75% of the wood fibre used by the European pulp industry in 2015 was in roundwood form, and the remaining 25% was made up of residual chips from European sawmills (CEPI, 2015). The supply of softwood residues increased slightly in the four years to 2015 because of increasing lumber production, and the share of total fibre consumption held by softwood residual chips increased from 22.6% in 2012 to 23.2% in 2015.

The biggest change in 2015 was the reduced reliance of the pulp industry on imported hardwood pulplogs and wood chips. In 2015, imported hardwood logs accounted for 31% of the total consumption of hardwood logs, and the share of imported softwood chips was just less than 19%; volumes of both have trended downward since 2009.

4.3 COMMONWEALTH OF INDEPENDENT STATES

4.3.1 Industrial roundwood markets

An estimated 210 million m^3 of industrial roundwood was harvested in the CIS subregion in 2015, the highest volume in at least 15 years (Table 4.3.1). Timber harvests in Belarus, the Russian Federation and Ukraine have increased steadily in the last few years, predominantly because of rises in exports of softwood lumber.

TABLE 4.3.1

Industrial roundwood balance, CIS, 2014-2016

(thousand m^3)

	2014	2015	2016f	Change (%) 2014-2015
Removals	208,029	210,258	216,512	1.1
Imports	559	550	550	-1.6
Exports	27,022	25,550	26,599	-5.4
Apparent consumption	181,566	185,259	190,463	2.0

Note: f = 2015 Committee on Forests and the Forest Industry forecast.
Source: UNECE/FAO, 2016.

The industrial roundwood harvest in the Russian Federation increased for the third consecutive year in 2015, to an estimated 190 million m^3, of which two-thirds were coniferous species. The increase in the timber harvest in 2015 was not uniform across the country, however: it was higher in the eastern provinces of Siberia and the Russian Far East, and harvests declined in most central and western provinces (WhatWood, 2016).

The accuracy of harvesting data in the Russian Federation is uncertain because, in addition to official estimates, the Russian government acknowledges that there is "undocumented" timber harvesting. Estimates of the volume of timber logged without permission vary substantially. In 2009, the Head of the Federal Forestry Agency in the Russian Federation reported that illegal logging may well lie in the range of 25-30 million m^3 annually. Recent government reports have estimated the illegally logged volume at 2 million m^3, but WWF Russia and the World Bank put the number at 35-40 million m^3. At a conference on timber management in 2013, President Putin indicated that illegal logging in the Russian Far East had increased by nearly 70% in the previous five years (President of Russia official website, 2013).

Domestic log consumption increased by 2.2% in the Russian Federation in 2015, to 171 million m^3. Higher demand for logs was driven mainly by improved export opportunities for processed products such as sawnwood due to a weakening rouble rather than to a rise in the domestic consumption of forest products.

4.3.2 Trade of roundwood

All three major forest-rich countries in the CIS – Belarus, the Russian Federation and Ukraine – export significant percentages of their timber harvests in unprocessed form. In 2015, log exports accounted for an estimated 13% of the harvest in the three countries combined. Export volumes were down, however, for both the Russian Federation and Ukraine because of slowing demand, especially in China, Finland, the Republic of Korea and Turkey.

The governments of Belarus and Ukraine have decided to follow in the footsteps of Canada, the Russian Federation and the US by banning or restricting the export of unprocessed logs in the hope of creating more local jobs by processing timber domestically. The log export ban by Ukraine (planned to be a EU member in 2020), has not been well received by the EU; the European Commission has made efforts to stop it since it is not in accordance with the liabilities of the Ukraine-EU Association Agreement, but negotiations have so far been in vain. According to a report by the European Parliament, Ukraine is in breach of the regulations of both the EU and the World Trade Organization by restricting free trade in the form of log exports (It's Ukraine, 2016). In March 2016, the Ukraine trade minister indicated that the export ban might be revoked and replaced with a log sales process that would prioritize domestic over foreign buyers (Interfax-Ukraine, 2016).

Under the existing Ukrainian policy, log exports from the country are banned for ten years as of 1 November 2015, with the exception of pine log exports, which will be banned from 2017. Of the estimated 2.7 million m^3 of softwood logs exported by Ukraine in 2015, almost 90% went to three countries: China, Romania and Turkey. Romania is the major destination, accounting for 47% of Ukraine's total shipments of softwood logs in 2015. Ukraine's log export ban could have severe consequences for the sawmill industry in Romania because Ukrainian logs previously accounted for about 15% of that country's total softwood sawlog needs.

In 2015, the President of Belarus signed a law banning log exports, starting 1 January 2016. The law contains a clause, however, declaring that it is possible to exempt some exporters from the ban, making it difficult to predict how the log trade will change as a result of the new law.

Source: UNECE/FAO, 2016.

The Belarusian Minister of Forestry stated that the domestic forest industry is unable to consume all harvested timber at the current level of production capacity, estimating that there was a surplus log volume of 2 million m^3. In 2015, Belarus shipped the majority of its softwood and hardwood export logs to the Baltic States, China, the Czech Republic and Poland.

4.4 NORTH AMERICA

4.4.1 Industrial roundwood markets

The total timber harvest in North America was almost unchanged in 2015 compared with 2014, at 506 million m^3, with coniferous and non-coniferous species accounting for 389 million m^3 and 117 million m^3, respectively.

According to official harvesting statistics, US removals in 2015 were almost the same as in 2011, at 355 million m^3 (Table 4.4.1). Expert analysis based on derived log consumption by the forest industry in the US and net log trade, however, indicates that actual removals of industrial roundwood were closer to 411 million m^3 in 2015, which was 10% higher than in 2011 (the totals in table 4.4.1 are based on official figures). The major reason for the increase in log consumption in the five-year period was higher lumber and pellet production. Softwood lumber production was 21% higher in 2015 than in 2011.

TABLE 4.4.1

Industrial roundwood balance, North America, 2014-2016

(thousand m^3)

	2014	2015	2016f	Change (%) 2014-2015
Removals	505,637	506,036	504,805	0.1
Imports	5,171	5,810	5,831	12.4
Exports	20,658	17,561	17,445	-15.0
Apparent consumption	490,150	494,286	493,191	0.8

Note: f = 2015 Committee on Forests and the Forest Industry forecast.
Source: UNECE/FAO, 2016.

The substantial rise in both softwood and hardwood lumber production in the US in recent years resulted in a higher percentage of timber removals processed in sawmills in 2015 than in 2011. Log consumption by the country's pulp mills and log exports both declined over the period.

Source: UNECE/FAO, 2016.

According to government-generated timber harvest statistics, log production and consumption both increased in Canada by about 3% in the five years to 2015, reaching 151 million m^3 and 150 million m^3, respectively. According to expert analysis based on derived domestic log consumption and net log trade, however, roundwood removals increased by 17% from 124 million m^3 in 2011 to a nine-year high of 145 million m^3 in 2015 (the totals in table 4.4.1 are based on official figures). Most of the increase in log consumption was because of higher production levels in Canada's softwood sawmills.

4.4.2 Trade of roundwood

Net log exports by North America fell from 15.5 million m^3 in 2014 to 11.8 million m^3 in 2015, due mainly to shrinking demand for softwood logs in Asia. Log exports from the US west coast to Asia fell substantially (by 33%) from 2013 to 2015, to 6 million m^3. The reduced demand for US logs was not limited to China; demand was also down in Japan and the Republic of Korea. A combination of lower demand for wood raw materials by sawmills in Asia and a strong US dollar meant that US log shipments fell to their lowest levels in five years.

Log exports from British Columbia to Asia have also fallen in recent years, from just over 6 million m^3 in 2013 to 4.9 million m^3 in 2015. Shipments declined to both the major markets, China (by 20%) and Japan (by 30%), in the two-year period; on the other hand, exports to the Republic of Korea have increased in recent years. A turnaround in trade could be detected in early 2016, with both China and Japan increasing their log-buying in western Canada. Exports to Japan were up by 24%, year-on-year, in the first five months of 2016.

Canada continues to be a net importer of softwood logs from the US, with most of the trade in the eastern part of the two countries. Canada's net imports were 2.2 million m^3 in 2015, down from 2.6 million m^3 in 2013.

4.5 EXTRAREGIONAL INFLUENCES AFFECTING THE UNECE REGION

The international trade of softwood logs was generally higher in the first few months of 2016 than in the same period in 2015. Three of the five-largest importing countries, Austria, China and Sweden, imported more logs in early 2016 than in the same period in 2015. From 2013 to 2015, however, eight of the top ten importing countries reduced their imports of softwood logs, with the biggest declines (in descending order, by volume) occurring in Japan, the Republic of Korea, Sweden and Canada.

China has been the world's largest importer of softwood logs for 15 years, due to its lack of domestic forest resources to meet the raw-material demands of its forest industry. Although China's import volume fell by 18% in 2015, the country still imported almost 40% of the softwood logs traded globally.

New Zealand and the Russian Federation continue to be the major suppliers of softwood logs to China, together accounting for almost 70% of all imports, followed by (in descending order, by volume) the US, Australia and Canada. Perhaps the most interesting development in the Chinese log market in the past few years has been Australia's increasing presence there. In three years, Australia increased its log shipments to China almost three-fold and now has a 10% market share, equal to that of the US. Over the past three years, Australian logs were consistently at the low end of the cost curve; in the first quarter of 2016, cost, insurance and freight (CIF) average prices for Australian logs were slightly lower than those for logs from the Russian Federation and 12% below the average price of all imported logs.

Source: UNECE/FAO, 2015.

The biggest year-on-year change in log exports in the first three months of 2016 was the decline in shipments from the US, Canada and New Zealand and the increase in export volumes by the Russian Federation, Norway and Australia.

4.6 WOOD RAW-MATERIAL COSTS

Wood fibre accounted for about 58% of the total production costs of the global pulp industry in the manufacture of pulp in 2015, and it was as high as 67% in some countries consuming plantation pulpwood (Fisher International, 2016). The wood raw-material cost for forest product manufacturers is not only the largest cost factor, it is also typically the most variable. Wood raw-material costs trended downward in much of 2015 and in early 2016, with sawlog and pulpwood prices reaching their lowest levels in more than six years in the first quarter of 2016.

4.6.1 Sawlog prices

Global softwood sawlog prices fell substantially between 2014 and 2016. The Global Sawlog Price Index, which is denoted in US dollars, fell to its lowest level since 2009 in the first quarter of 2016 (graph 4.6.1). The index has fallen 20.7% since the first quarter of 2014 due to a strengthening US dollar and lower demand for lumber in Asia, the Middle East and North Africa, and Europe.

GRAPH 4.6.1

Global Sawlog Price Index, softwoods, 2006-2016

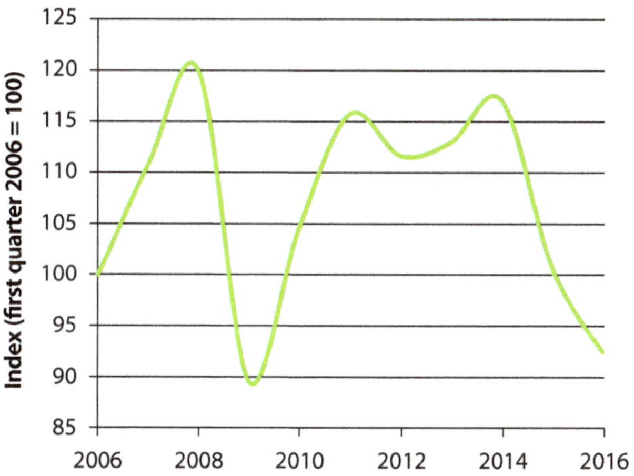

Note: Price index based on delivered sawlog prices in 19 key regions worldwide.
Source: Wood Resource International, 2016b.

Sawlog prices have trended downward in a number of European countries in recent years in both US dollars and local currencies (graph 4.6.2). The European Sawlog Price Index, which represents nine major softwood sawlog markets in Europe, declined by 10% in euro terms from the first quarter of 2014 to the first quarter of 2016 (Wood Resource International, 2016b). The biggest price declines in recent years have been in Austria, the Czech Republic and Germany.

GRAPH 4.6.2

Softwood sawlog price indices in selected countries, 2011-2016

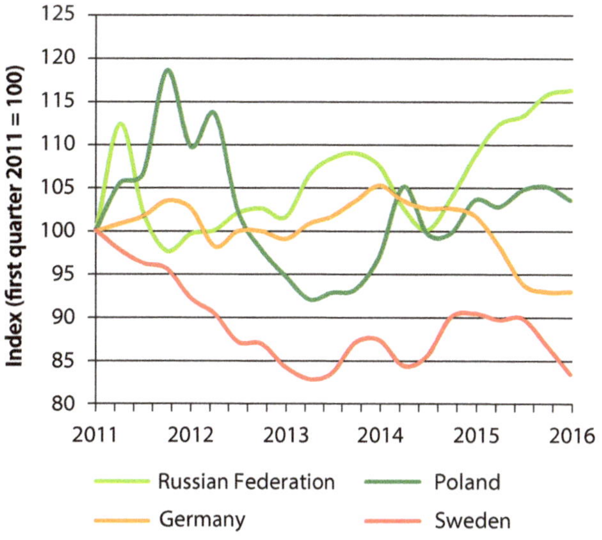

Note: Price indices based on delivered log price per m³ in local currencies.
Source: Wood Resource International, 2016b.

Despite recent substantial log-price reductions, sawmills in Austria and Germany still have higher wood costs than lumber producers in the Nordic countries and eastern Europe. High costs for domestically sourced sawlogs have driven sawmills in both Austria and Germany to increasingly source wood raw materials from neighbouring countries, where log prices are lower.

The continued weakening of the rouble in 2015 and early 2016 resulted in a substantial reduction in US dollar domestic sawlog prices in the Russian Federation. Average domestic log prices in the Russian northwest are down by 50% compared with two years ago, and softwood sawmills there enjoy some of the world's lowest wood raw-material costs.

In North America, price movements in the last few years have been mixed, with US-dollar prices declining in Canada and the US West and almost unchanged in the US South (graph 4.6.3). Log prices fell in Canada because of the stronger US dollar[13] and in the US West, where prices in the first quarter of 2016 were 13% below those recorded in the first quarter of 2014; weakening log demand was the main cause of downward price pressure.

[13] The Canadian price decline is only due to US dollar appreciation; it is thus not shown in graph 4.6.3, which is based on local currencies.

GRAPH 4.6.3

Softwood sawlog price indices in North America, 2011-2016

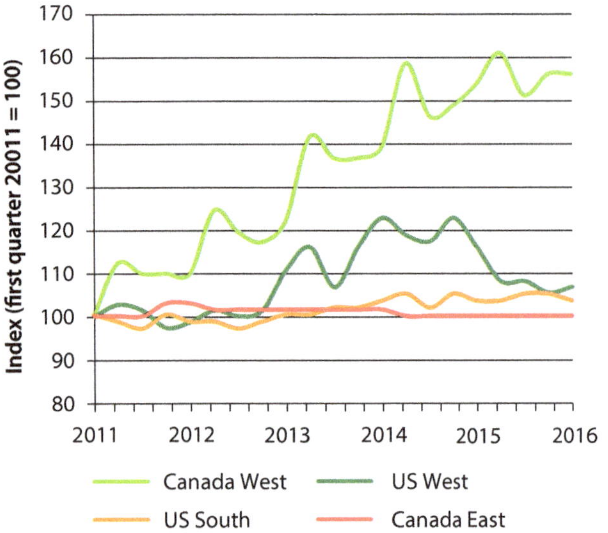

Notes: Price indices based on delivered log price per m³ in local currencies. "US South" indicates prices for pine; "US West" indicates prices for hemlock; "Canada West" and "Canada East" indicate prices for mixed conifers.
Source: Wood Resources International, 2016a.

4.6.2 Pulpwood prices

Wood costs in the pulp industry have fallen steadily worldwide in US dollar terms for almost five years, with hardwood fibre prices declining most (Wood Resource International, 2016b). The Hardwood Wood Fiber Price Index has declined consistently in the last few years; in the first quarter of 2016 it was 29% lower than its record high in 2011 (graph 4.6.4). The biggest price drops in hardwood fibre since the 2011 peak have been mainly in countries with fast-growing plantations, such as Australia, Brazil, Chile and Indonesia. The price declines have been entirely the result of the stronger US dollar, with hardwood log costs in local currencies increasing by more than 25% in countries such as Brazil, Chile, Indonesia and the Russian Federation.

GRAPH 4.6.4

Softwood Wood Fiber Price Index and Hardwood Wood Fiber Price Index, 1990-2016

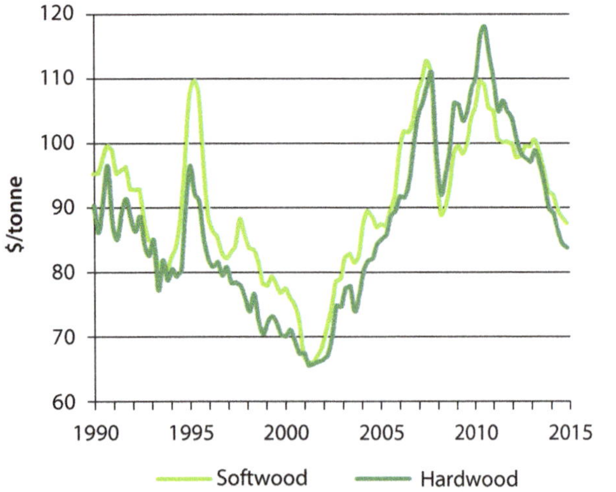

Notes: Price indices are based on delivered pulpwood and wood chip prices in 18 key regions worldwide. Indices are based on US dollars per oven-dry tonne.
Source: Wood Resource International, 2016b.

Softwood fibre prices fell in a number of countries worldwide in the first quarter of 2016, in both US dollars and local currencies (graphs 4.6.5 and 4.6.6). The downward price trend that started in Europe in early 2014 continued, but prices are starting to level out in some markets and the bottom may have been reached. The biggest price declines from the first quarter of 2015 to the first quarter of 2016 were in Brazil, Canada, France, New Zealand and the Russian Federation.

Source: E. O'Driscoll, 2015.

The Softwood Wood Fiber Price Index dropped by 5% from the first quarter of 2015 to the first quarter of 2016, reaching its lowest level since the first quarter of 2006. The only region where wood fibre costs have increased in the last few years is the US South, where average softwood pulplog prices in the first quarter of 2016 were 21% higher than in 2012.

Wood fibre costs for European pulp mills continued to decline in 2015. There has been a steady and substantial reduction in fibre costs throughout Europe in the last 2-3 years. In the Nordic countries, prices for chips and pulplogs have fallen by 20-25% (US dollar terms) in the last two years. Pulplog prices have declined by 25-40% in central Europe since 2013 and by 18-28% in Iberia (Portugal/Spain).

Many pulp mills in Europe, which are benefiting from the lowest wood fibre costs for more than eight years, have become more competitive with pulp companies in Canada and the US. Although the price discrepancy between North America and Europe has diminished in recent years, however, especially for hardwood fibre, there are still substantial price differences between the two subregions for softwood fibre, favouring North America.

GRAPH 4.6.5

Softwood chip price indices in selected countries, 2011-2016

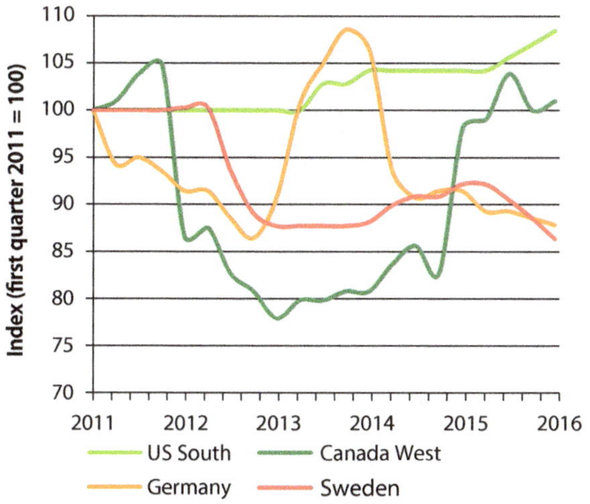

Note: Price indices are based on delivered wood chip price per oven-dry tonne in local currency.
Source: Wood Resource International, 2016b.

GRAPH 4.6.6

Softwood pulplog price indices in selected countries, 2011-2016

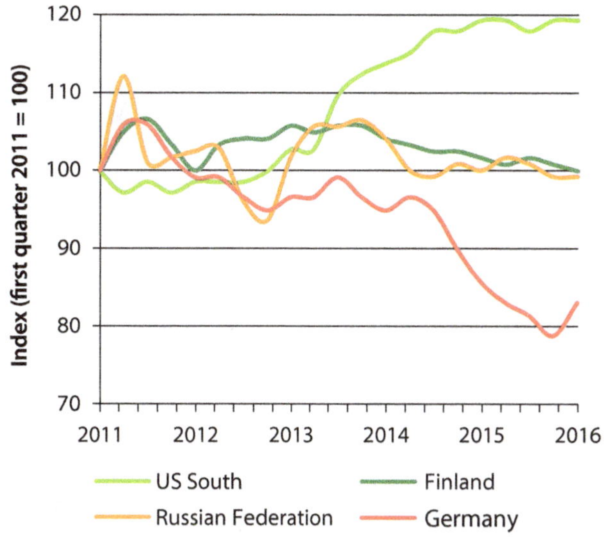

Note: Price indices are based on delivered log price per oven-dry tonne in local currency.
Source: Wood Resource International, 2016b.

Note: The statistical annex of the Forest Products Annual Market Review 2015-2016 is available at: www.unece.org/forests/fpamr2016-annex

4.7 REFERENCES

CEPI. 2016. Confederation of European Paper Industries (CEPI). Available at: www.cepi.org

Fisher International. 2016. Available: at www.fisheri.com

Interfax-Ukraine. 2016. Cabinet proposes lifting ban on timber exports, introducing timber auctions (25.03.2016). Available at: http://en.interfax.com.ua/news/general/333252.html

It's Ukraine. 2016. From the EU are signals that Ukraine is in violation of the Agreement on Association — sales representative (14.04.2016). Available at: http://en.reporter-ua.ru/from-the-eu-are-signals-that-ukraine-is-in-violation-of-the-agreement-on-association-sales-representative.html

President of Russia official website. 2013. State Council Presidium meeting on strengthening the timber sector (11.04.2013). Available at: http://en.special.kremlin.ru/events/president/news/17876

UNECE/FAO. 2016. TIMBER database. Available at: www.unece.org/trade/timber

WhatWood. 2016. Available at: www.whatwood.ru

Wood Resources International. 2016a. North American wood fiber review. Available at: www.woodprices.com

Wood Resources International. 2016b. Wood resource quarterly. Available at: www.woodprices.com

5 SAWN SOFTWOOD

Lead author: ***Russ Taylor***

Contributing authors:

Antti Koskinen,
Fran Maplesden
and Igor Novoselov

HIGHLIGHTS

- Apparent sawn softwood consumption rose by 6.1% in North America in 2015, was flat in Europe, and decreased in the CIS subregion by 2.2%.
- Although overall apparent sawn softwood consumption was steady in Europe in 2015 (at 89 million m^3), there were widely varying changes in consumption among countries. Germany remained the largest market in the European subregion. Turkey has quickly become the subregion's fourth-largest consumer of sawn softwood.
- Sawn softwood production decreased in Europe by 0.7% in 2015, to 102.6 million m^3. Notable gains were in Austria, Poland, Sweden and Turkey, which collectively added 1.4 million m^3.
- Growth in European sawn softwood exports levelled off at 1% (with a total volume of 48.9 million m^3) in 2015, compared with 5% in 2014. EU28 exports to overseas markets declined by 1%, to 20.0 million m^3, but intrasubregional exports grew slightly.
- Difficulties for European sawn softwood exporters in the key markets of Egypt (down by 14% in 2015) and Japan (down by 5%) were compensated by success in other markets. Sawn softwood exports to China grew by 20% in 2015, to 2.1 million m^3, and exports to the US grew by 44%, to 0.4 million m^3.
- Sawn softwood production in the Russian Federation increased by 0.2% in 2015, to 32.1 million m^3. Larger export-oriented mills were able to maintain or increase production, but many smaller mills selling into the domestic market decreased their output.
- US housing starts (number of new housing units initiated) reached 1.11 million units in 2015 (up by 11% compared with 2014), with multifamily housing starts reaching a record high. Multifamily construction consumes about 65% less wood per family unit than do traditional single family units.
- Sawn softwood production in Canada and the US grew by 8.3% and 1.1%, respectively, in 2015.
- Sawnwood prices (in US dollars) declined by 20% in North America in the first three quarters of 2015 before recovering in the fourth quarter and regaining most of the losses by mid-2016.
- A slowdown in China's construction market, coupled with increased Russian and European sawn softwood exports, had a negative impact on exports from Canada and the US. Canadian exports dropped by 15% in 2015 compared with 2014, and US exports plunged by 30%. Canadian exports declined by a further 3% in the first five months of 2016, but US exports increased by 8%.
- The latest US–Canada Softwood Lumber Agreement expired in mid-October 2015. It is expected that the US Department of Commerce will be asked to investigate Canadian lumber imports, which could result in the imposition of punitive duties on Canadian lumber exports to the US by March 2017.

5.1 INTRODUCTION

As in 2014, 2015 saw generally mixed and unsettled global economic trends. The recovery in North America continued for the sixth consecutive year, and sawn softwood consumption increased by 6.1%. Sawnwood consumption was steady in Europe, but economic conditions and depreciating currencies in the CIS countries resulted in a decline in sawn softwood consumption of 2.2% in that subregion (table 5.1.1). The US dollar strengthened against most currencies in the first half of 2015, and volatile exchange rates affected countries differently. Sawn softwood production increased in North America by 4.2% in 2015 and in the CIS by 0.4%, while output in Europe declined by 0.7%.

TABLE 5.1.1

Apparent consumption of sawn softwood in the UNECE region, by subregion, 2014-2015

(thousand m³)

	2014	2015	m³/capita (2015)	Change (%) 2014-2015
Europe	88,996	88,983	143.8	0.0
CIS	17,039	16,659	60.6	-2.2
North America	85,506	90,746	260.3	6.1
Total	191,541	196,388	174.0	2.5

Source: UNECE/FAO, 2016.

5.2 EUROPE

4.2.1 5.2.1 Consumption

The optimism and anticipated growth forecast a year ago levelled off in 2015, with apparent consumption in Europe remaining at the 2014 level of 89 million m³ (table 5.2.1). Economic conditions are highly variable between countries in the subregion, and this is reflected in the wide range of growth rates in sawn softwood consumption in 2015; as in 2014, there were both underperforming markets and others that showed healthy growth.

TABLE 5.2.1

Sawn softwood balance, Europe, 2014-2016

(thousand m³)

	2014	2015	2016f	Change (%) 2014-2015
Production	103,246	102,565	102,671	-0.7
Imports	34,207	35,338	34,717	3.3
Exports	48,458	48,920	50,761	1.0
Apparent consumption	88,996	88,983	86,627	0.0

Note: f = 2015 Committee on Forests and the Forest Industry forecast.
Source: UNECE/FAO, 2016.

Only five of the ten highest sawn softwood-consuming countries in the European subregion increased consumption in 2015; combined, the top ten countries returned a negative result. The highlight in 2015 was Turkey, where consumption increased by 7.2%, year-on-year, to 5.6 million m³; this country is now the fourth-largest market in the subregion. The Central European market also reported healthy growth, with Germany (+1.5%), Austria (+4.2%) and Poland (+4.2%) all increasing consumption; in Austria, this was after several years of decline.

Source: UNECE, 2015.

Belgium and Finland were the most disappointing sawn softwood markets in Europe in 2015. Both reported double-digit percent declines in consumption; combined, their consumption dropped by 1.0 million m³. GDP growth in both countries is well below the EU average. In Finland, the increasing popularity of wooden high-rise construction has been insufficient to compensate for the eroding market in single-family homes, of which the number of starts in 2015 was the lowest this century. France and the UK – the two largest importers of sawn softwood in Europe – both reported declines in consumption in 2015. This was the fourth consecutive year of decline in France; on the other hand, consumption in the UK had been growing rapidly in recent years.

Germany is still the largest consumer of sawn softwood in Europe, at 18.5 million m³ in 2015, nearly twice the consumption of the UK, the subregion's second-largest consumer. Although traditional markets are struggling, there are some new, fast-growing markets. Estonia is still a relatively small market (consuming 1.7 million m³ in 2015), but it has the subregion's highest per capita sawn softwood consumption, due largely to its strong export-oriented remanufacturing sector, which requires sizeable imports of sawn softwood.

5.2.2 Production and capacity change

Europe produced 102.6 million m³ of sawn softwood in 2015, a small decrease (0.7%) over 2014. Producers lacked an incentive to increase production, given sluggish demand in Europe and key overseas export markets. Nevertheless, there were notable production increases in Austria, Poland, Sweden and Turkey, which collectively added 1.4 million m³ to total production.

Reported increases in these countries were between 3.9% and 5% and were due mainly to positive developments in domestic markets, although production growth in Sweden was driven by increasing exports.

There were no clear production trends in Europe in 2015. Output increased in Sweden but decreased by 2.6% in neighbouring Finland. Production went down by 0.3 million m^3 in Germany but increased in Austria by a similar volume. Other significant production decreases occurred in France (2.3%), Slovakia (-3.4%), Romania (-5.5%) and the UK (-7.2%). Production has been decreasing for several years in France, but it had increased in the UK for several years before 2015.

Source: UNECE, 2016.

Production capacity in Europe is still sufficient in the prevailing market conditions, and there were no major changes in capacity in 2015 and the first half of 2016. Some minor closures took place in Central Europe, and existing mills continue to upgrade to remove bottlenecks, thereby increasing productivity and production capacity.

5.2.3 Prices

Prices for European sawn softwood varied by market in 2015 (graph 5.2.1). In Germany they were relatively stable, with a nominal decrease of 1.4% (in euros per m^3); there was almost no price movement in the first half of 2016.

In the Middle East, CIF prices increased by 7.6% in 2015 over 2014. This was due to higher prices in the first half of the year, but prices declined thereafter, flattening to a level similar to that in 2014. As in Germany, prices for European sawn softwood exports to the Middle East hardly changed in the first half of 2016, despite increasing volumes.

Prices in Japan for European sawn softwood decreased in the local currency in 2015, but recent price developments have been in favour of European exporters due to the weakening of the euro against the yen. Free-on-board (FOB truck Japanese port) prices dropped in euro terms after the first half of 2015, and the average price in 2015 was 2.1% lower than in 2014. Nevertheless, the depreciation of the euro against the yen in 2016 means that prices in euros were 13% higher in mid-2016 than in June 2014 and at their highest since 2012.

GRAPH 5.2.1

European sawn softwood prices in Germany, Japan and the Middle East, 2010-2016

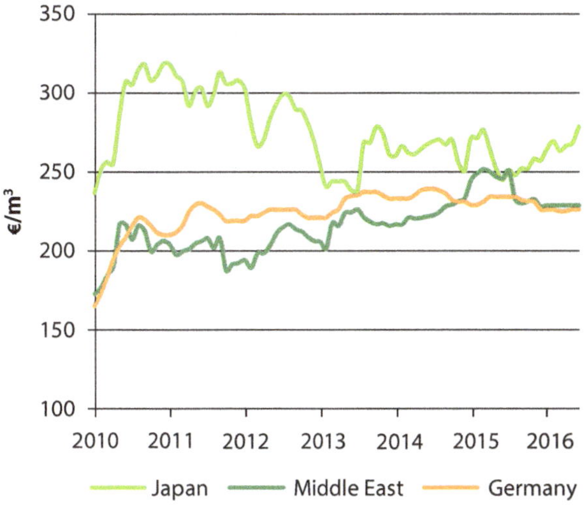

Note: Data to June 2016. Japan: European whitewood Lamina, KD Rough FOB truck port yard. Germany: roof framing lumber, delivered. Middle East: Scandinavian/Baltic whitewood and red pine, sixths, CIF.
Source: EUWID Wood Products and Panels, 2010-2016; Japan Lumber Report, 2012-2016; Wood Markets Monthly, 2010-2016.

5.2.4 Trade

5.2.4.1 Imports

European imports of sawn softwood grew in 2015, with half this growth credited to imports originating outside the subregion. The import volume was 35.3 million m^3 in 2015, up by 3.3% compared with 2014. Around 80% of this volume was intrasubregional, but extrasubregional imports increased by 9%, reaching 6.8 million m^3 in 2015. Russian sawn softwood exports to Europe benefited from the devaluation of the rouble, with the EU28 importing 3.3 million m^3 from the Russian Federation in 2015, the highest volume since 2010. Imports from Belarus grew by 29%, to 1.0 million m^3.

Europe's sawn softwood imports from North America halved in 2015, to 0.3 million m^3, with US sawmills finding the domestic market more attractive. On the other hand, imports from the "plantation pine countries" (Brazil, Chile and New Zealand) nearly tripled in 2015, to 0.2 million m^3, with pine clears increasingly used in European remanufacturing.

5.2.4.2 Exports

Growth in European sawn softwood exports levelled off in 2015, to 1% (compared with 5% in 2014), at a volume of 48.9 million m^3. EU28 exports to overseas markets declined by 1% in 2015, to 20.0 million m^3, and overall (albeit minimal) growth was achieved thanks to intrasubregional exports. Demand for European sawn softwood declined in the two main export markets, Japan and Egypt, by 5% and 14%, respectively. In Japan, the decline was clearly related to an overall decline in demand; the overall market was stable in Egypt, but Russian exporters gained market share there due to a currency-exchange advantage.

The difficulties in sawn softwood exports to Japan and Egypt were compensated by success in other markets. European exports to China continue to grow rapidly, increasing by 20% in 2015, to 2.1 million m³. The US market is also a growing market, with exports up by 44% in 2015, to 0.4 million m³. The overall situation in overseas markets is problematic, especially for Nordic sawmills, with major differences in the supply–demand balance for spruce and pine. Nordic spruce is in strong demand in China and the US, but pine is struggling, with exports decreasing, especially to Egypt.

First-quarter 2016 export data indicate positive developments, with European exports to Japan and North Africa up by 33% and 7%, respectively. For example, exports have increased from Finland and Sweden to Egypt, which should ease pressure in sawn pine demand there. Exports to China by Finland and Sweden also grew in the first quarter of 2016, by 66% and 33%, respectively.

5.3 COMMONWEALTH OF INDEPENDENT STATES, WITH A FOCUS ON THE RUSSIAN FEDERATION

5.3.1 Consumption

Apparent sawn softwood consumption decreased by 2.2% in the CIS subregion in 2015, to 16.7 million m³ (table 5.3.1).

5.3.2 Production/capacity change

The production of sawn softwood in the CIS subregion was estimated at 36.3 million m³ in 2015, up by 0.4% from 2014.

Most Russian sawmills sought to export products in 2015. The devaluation of the rouble enabled them to achieve large sales margins and high profitability, despite the strong decline in global prices (in US dollars) in the key sawn softwood export markets. Rising prices for logs and other costs, and low sale prices, constrained sales and output in the Russian domestic market.

Source: R. Vlosky, 2014.

Russian sawn softwood production increased slightly (by 0.2%) in 2015, to 32.1 million m³, but domestic consumption fell by 9%, to 9.8 million m³. The drivers of production growth were in the Irkutsk region (+4%, year-on-year) and the Arkhangelsk region (+8%); these two regions accounted for 26% of the total Russian production volume in 2015, and the Krasnoyarsk region (2% growth in 2015) accounted for 11%. Most large export-oriented mills were able to maintain or increase production volumes, but many small sawmills selling into the domestic market decreased their output.

TABLE 5.3.1

Sawn softwood balance, CIS subregion, 2014-2016

(thousand m³)

	2014	2015	2016f	Change (%) 2014-2015
Production	36,178	36,314	36,613	0.4
Imports	4,764	5,060	5,060	6.2
Exports	23,902	24,714	24,714	3.4
Apparent consumption	17,039	16,659	16,959	-2.2

Note: f = 2015 Committee on Forests and the Forest Industry forecast.
Source: UNECE/FAO, 2016.

The acquisition by the Segezha Group (part of the conglomerate company, AFK Sistema) of Lesosibirsk LDK №1 (Krasnoyarsk region) was a significant development in the Russian sawmill industry in 2015. On completion of the merger, the Segezha Group will be one of the largest sawnwood producers in the Russian Federation (around 1 million m³ per year).

5.3.3 Prices

According to Rosstat (2016), the weighted average price for Russian sawn softwood producers in 2015 was 5,501 roubles per m³ (US$90 per m³) in the domestic market (down by 6.5%, year-on-year) and 10,607 roubles per m³ (US$175 per m³) in export markets (up by 28.7%, year-on-year) (graph 5.3.1).

The dynamics of domestic prices in the Russian Federation were contrary to inflation trends in 2015. The growth of export prices was due entirely to the weakening of the national currency.

GRAPH 5.3.1

Sawn softwood prices in the Russian Federation, 2010-2016

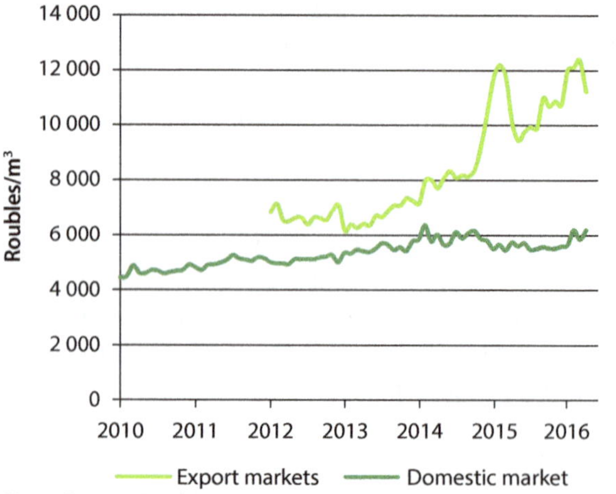

Note: Data to April 2016.
Source: Rosstat, 2016.

5.3.4 Trade

Traditionally, exports of Russian sawn softwood are mainly in the form of rough sawnwood (96% of the total volume in 2015, at 21.5 million m³), of which rough-sawn pine accounts for more than 60%.

The volume of sawn softwood exports from the Russian Federation achieved a record high of 22.4 million m³ in 2015, up by 5% from 2014. China consolidated its position as a key market for Russian sawn softwood (graph 5.3.2). According to WhatWood (2016), China's share of Russian exports rose by 5% in 2015, to 44% of the total volume; China imported 9.8 million m³ of Russian sawn softwood, valued at $1.17 billion (note, however, that estimates according to China Customs data differ from those generated from Russian export data). Strong growth in sawn softwood purchases from the Russian Federation in 2015 were also observed in Egypt (up by 33%, to 2.0 million m³), Iran (up by 18%, to 779,000 m³), the Republic of Korea (up by 41%, to 385,000 m³), and the UK (up by 18%, to 373,000 m³).

There was moderate growth in shipments of Russian sawn softwood to Estonia (+4.1%, to 538,000 m³), Germany (+5.8%, to 445,000 m³), and Japan (+1.8%, to 847,000 m³). There was a sharp drop in exports to Uzbekistan (-14%, to 2.4 million m³), Italy (-22%, to 92,500 m³), and France (-7.1%, to 283,000 m³).

GRAPH 5.3.2

Russian Federation sawn softwood exports by market, 2015

(million m³)

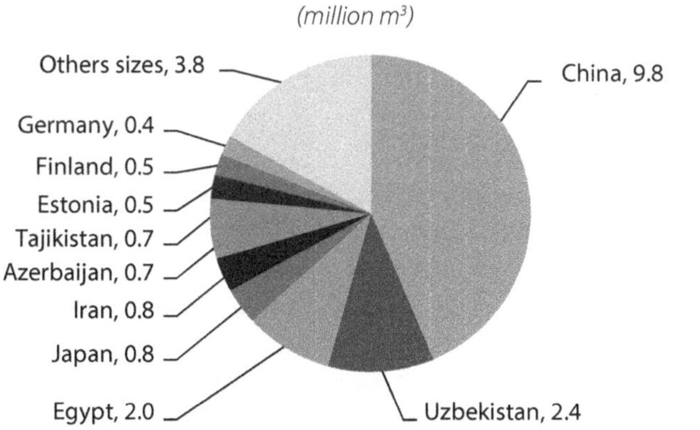

Note: Excludes trade with Eurasian Customs Union[14].
Source: WhatWood, 2016.

Russian exports of sawn softwood to European countries grew by 1% in 2015, to 3.26 million m³. The largest increase was to Latvia, where imports of Russian sawn softwood were up by 98%, to 202,000 m³. On the other hand, Russian exports declined to Finland (down by 35%, to 283,000 m³) and Italy (down by 22%, to 92,000 m³) (graph 5.3.3).

GRAPH 5.3.3

Russian Federation sawn softwood exports to Europe, 2014 and 2015

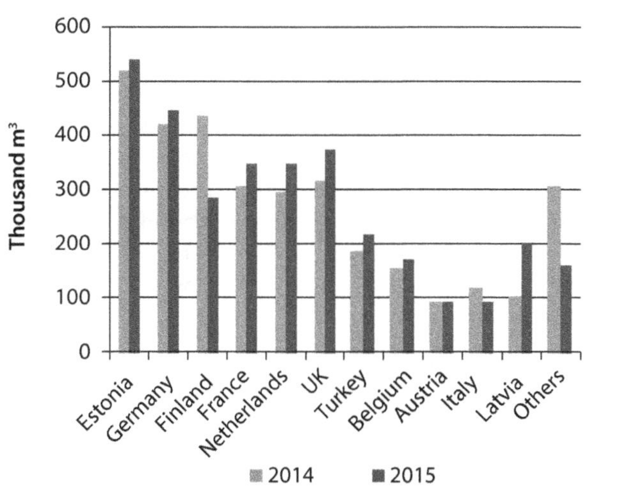

Source: WhatWood, 2016.

Siberian pine is the most popular export species from the Russian Federation, constituting 63% of the export volume (14 million m³) in 2015 (graph 5.3.4). The market share of larch increased from 8% in 2009 to 10% in 2015 and now accounts for 2.2 million m³ of exports.

GRAPH 5.3.4

Russian Federation sawn softwood exports by species, 2015

(million m³)

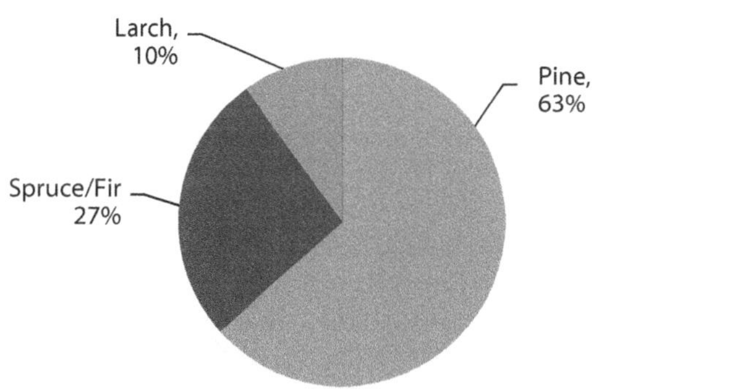

Source: WhatWood, 2016.

5.4 NORTH AMERICA

5.4.1 Consumption

Demand in North American sawn softwood markets increased in 2015 and the first half of 2016. The primary driver of consumption continues to be the rebound in US housing starts, which reached 1.11 million units in 2015 (up by 11% compared with 2014) (US Department of Census, 2016). There were 397,300 multifamily housing starts in 2015, the highest number since 1989. Notably, multifamily construction consumes approximately 65% less sawn softwood and wood-based panels per family unit than do traditional single-family units. Industry-based promotional efforts, such as the Softwood

[14] The Eurasian Customs Union has five member states, Armenia, Belarus, Kazakhstan, Kyrgyzstan, the Russian Federation and Tajikistan (prospective member).

Lumber Board (SLB)[15] initiative to increase wood use (including cross-laminated timber) in taller/larger apartment buildings, should lead to further increases in North American wood consumption.

The US economic outlook is healthy, and GDP growth should remain within the ideal range of 2-3% per year. Growth in residential housing starts, continued strength in repair and remodelling activity in both the US and Canada, and gains in the non-residential sector, drove a 6.1% increase in apparent North American sawn softwood consumption in 2015 (table 5.4.1). Of the total consumption of 90.8 million m³, 75.0 million m³ (up by 4.2%, year-on-year) was in the US and 15.8 million m³ (up by 16.6%) was in Canada.

TABLE 5.4.1
Sawn softwood balance, North America, 2014-2016
(thousand m³)

	2014	2015	2016f	Change (%) 2014-2015
Production	95,694	99,695	101,725	4.2
Imports	21,796	24,059	24,165	10.4
Exports	31,984	33,008	34,317	3.2
Apparent consumption	85,506	90,746	91,574	6.1

Note: f = 2015 Committee on Forests and the Forest Industry forecast.
Source: UNECE/FAO, 2016.

5.4.2 Production and capacity change

US sawn softwood output was 54.3 million m³ in 2015, an increase of 1.0% compared with 2014. Production gains were highest in the Midwest and Northeast regions, followed by the Inland (5.0%) and the South (3.0%); the West Coast region recorded a decline of 2.5%. Ongoing depressed sawlog prices (since 2009) due to excess log supplies coupled with strong housing demand put the South – which accounts for more than 50% of US production – in a leading position among US sawn softwood-producing regions. Three Canadian companies (Canfor, Interfor and West Fraser) have been purchasing sawmill companies in the US South since 2007; combined, they now operate 36 sawmills there and represent 24% of the South's capacity.

Source: UNECE, 2015.

[15] The SLB is an industry-funded initiative established to promote the benefits and uses of softwood lumber products in outdoor, residential and non-residential construction.

Canadian sawn softwood production soared by 8.3% in 2015, to 45.4 million m³. The British Columbia Interior, Canada's leading region for sawn softwood production (accounting for 44% of national production in 2014), posted its largest gain – 5.4% – since 2011 (Statistics Canada, 2016). Two mills in the province's interior that exploded in 2012 due to excess sawdust caused by the processing of dead logs (from the mountain pine beetle epidemic) were in full operation in 2015. Nevertheless, a gradual reduction in mill shifts continued in the British Columbia Interior in 2015 as the pine sawlog supply dwindled and it was less economically feasible to mill dead standing trees. Expansions in existing mills in Alberta, Manitoba and Saskatchewan led to an increase in output of 9% in those three provinces in 2015.

Sawn softwood production surged by 16% in Eastern Canada (dominated by New Brunswick, Nova Scotia, Ontario and Quebec) in 2015 (Statistics Canada, 2016). Quebec remained the highest-producing province, accounting for 60% of the area's output, but the increase in sawn softwood production in Ontario (33%) in 2015 – due to two new sawmill start-ups and to other capacity increases – was the largest gain in North America.

The Canadian dollar has been depreciating against the US dollar since 2014, providing Canadian producers with a competitive advantage in shipments to the US market. This is one of the reasons why the increase in Canadian production was much higher than in the US in 2015. Another reason was that export duties on Canadian lumber to the US were eliminated for a one-year period, starting mid-October 2015, pending the development of a new softwood lumber agreement some time in 2017.

5.4.3 Prices

US dollar prices in major markets eroded from early 2014 to the middle or end of 2015. Overall market demand was relatively balanced with supply in 2015, and prices moved higher in the second half of 2015 and early 2016. Much of the price decline in the first half of 2015 was due to ongoing currency devaluations against the US dollar, with exporters giving away their currency exchange gains by lowering US dollar prices.

Chinese demand was higher in late 2015, and sawn softwood imports increased by 17% in the first five months of 2016. In Japan, domestic production continues to expand at the expense of exporters in a push for fibre self-sufficiency.

Europe's market share of softwood imports in Japan eclipsed that of North America for the first time in 2013. In 2015, Europe had a 41% share of Japanese sawn softwood imports, compared with North America's 39%.

The bellwether structural framing lumber composite price in the US fell by 20% in the first three quarters of 2015 and then increased by 20% in the next three quarters to mid-2016 (Random Lengths, 2016) (graph 5.4.1). Favourable demand forecasts in the US for the rest of 2016, coupled with tightening supply factors and the potential for punitive duties on Canadian lumber exports from the second quarter of 2017, suggest that US lumber prices will move higher.

GRAPH 5.4.1

Quarterly prices for sawn softwood in China, Europe, Japan and the US, 2005-2016

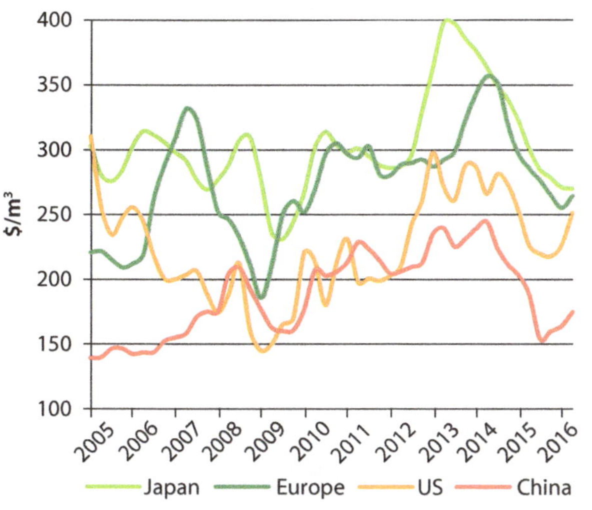

Note: Data to June 2016, delivered-to-market prices. Japan: BC W-SPF 2x4, J-grade, C&F; Europe: Swedish spruce 47x100, C&F; US: W-SPF grade #2&Btr, 2x4, delivered to Chicago; China: SPF/Hem-Fir, green, grade #3&Btr 1-7/8x4-12, C&F.
Source: Wood Markets Monthly, 2005-2016; Wood Markets China Bulletin, 2005-2016.

5.4.4 Trade

The end of the nine-year US–Canada Softwood Lumber Agreement in mid-October 2015 resulted in a one-year window of duty-free Canadian lumber exports to the US. Slightly improved conditions in most export markets from the third quarter of 2015 enabled US and Canadian sawn softwood producers to increase offshore exports to Asia and to take advantage of growing demand in North America.

The progressive strengthening of the US dollar ended in mid-2015, when US dollar prices started to bottom out. Because the Russian Federation had the largest currency devaluation of all exporting countries and offered very competitive prices, it gained market share in China at the expense of North American sawn softwood. Improving demand in China and Japan led to some gains in North American exports, but this trend started to slow in the second quarter of 2016.

A slowdown in China's construction market, coupled with a rise in Russian log and sawn softwood exports, had a negative impact on structural lumber exports from Canada and the US in 2015. Canadian exports dropped by 15% compared with 2014, to 5.6 million m³, and US exports plunged by 30%, to 590,000 m³. Canadian exports fell by a further 3% in the first five months of 2016, but US exports were up by 8%. Overall, sawn softwood exports to China shrank by 1.0% in 2015, to 17.5 million m³, with the Russian Federation the single-largest supplier, at 8.4 million m³.

Much of the decline in Japan's sawn softwood imports can be attributed to the rise of domestic sawn softwood output using both domestically produced and imported logs. Japanese sawnwood imports from all countries shrank by 4% in 2015, to 6.2 million m³. North American exports to Japan declined by only 1% (to 2.34 million m³).

5.4.4.1 Imports

Canada continues to dominate US imports, with a near 96% share in 2015. Canadian shipments to the US were up by 1.82 million m³ (9.2%) in 2015, to 21.75 million m³, the highest volume since the global financial crisis.

5.4.4.2 Exports

US sawn softwood exports shrank by 11% in 2015, to 2.6 million m³. The most significant reductions were to China (30%) and Asian countries other than China and Japan (22%).

Canadian sawn softwood exports to overseas markets declined by 5% in 2015, to 7.7 million m³. The largest decrease was to China but, despite this, China still accounted for 60% of Canada's overseas sawn softwood exports in 2015.

Source: UNECE, 2015.

Canadian sawn softwood exports to the US soared by 38% in the first three months of 2016 compared with the corresponding period in 2015, with exports to most other markets unchanged due to flat demand. With no export duties in place, growing demand and moderate prices, Canadian sawn softwood producers took full advantage of the strong US market.

The continuation of positive economic drivers and the potential for a tightening of the supply–demand balance suggest an optimistic outlook for growth in sawn softwood markets through to the end of 2016. For North American producers, the key metrics to watch are rising US consumption (housing starts), currency rates, export markets, and the outcome of US–Canada Softwood Lumber Agreement negotiations.

5.5 EXTRAREGIONAL INFLUENCES AFFECTING THE UNECE REGION

Outside the UNECE region, China continued to dominate sawn softwood imports, at 17.6 million m³ in 2015 (China Customs data) (table 5.5.1). Although this volume was up slightly (by 1%) year-on-year, its value was down 12%, reflecting a significant fall in import prices in 2015 (World Resources Institute, 2016) as construction demand slowed and the market became oversupplied. China's sawn softwood imports are destined mainly for use in housing and construction, and the slowdown in China's economy since 2014 (particularly the volatility in the housing sector) has had a dampening effect on demand for construction materials, although the effects have not been as large as expected. The building boom between 2011 and 2013 resulted in an oversupply of housing, and debate is ongoing about whether there is too much inventory for a sustainable increase in housing starts and whether the unsold housing inventory will normalize. China's imports in 2015 were predominantly from UNECE sources, particularly the Russian Federation (48%) and Canada (32%) but also Finland (3.5%), the US (3.4%), Sweden (3.0%) and Germany (1.1%). The only significant competitors outside the UNECE region were Argentina, Brazil, Chile and New Zealand, which together supplied 7% of the total sawn softwood import volume. China's imports from the Russian Federation increased by 16.5% (by volume) in 2015 compared with 2014. This trend has continued, with import volumes surging by 28% in the first five months of 2016, year-on-year.

Japan's demand for imported sawn softwoods continued to slow in 2015. Although new housing starts were up slightly (by 1.9%) compared with 2014, sawn softwood imports declined as domestic supply increased its share of total consumption (ITTO, 2016). The Japan Forestry Agency projected a decline in sawn softwood imports in 2016 because of falling production in the country's pre-cutting plants in response to slow demand and excess inventories (ITTO, 2016). New housing starts grew by 10% from January to April 2016 (well above forecasts made earlier in the year), however, in response to the introduction of financial support packages for home buyers, particularly low-interest mortgages (ITTO, 2016); consequently, sawn softwood imports rose by 7% (by volume) in the first four months of 2016, year-on-year. Imports have been assisted by the appreciation of the yen since November 2015, which has put downward pressure on import prices. In the medium term, however, housing starts are expected to contract in response to an expected decline in the number of households, with housing construction limited to rebuilding ageing infrastructure. Japan's medium-to-long-term economic outlook is weak, primarily reflecting a declining labour force (IMF, 2016). A scheduled increase in the consumption tax in April 2017, which was expected to dampen sawnwood demand, has been postponed to October 2019 amid concerns that an increase may trigger another economic recession.

TABLE 5.5.1

Major importers and exporters of sawn softwood outside the UNECE region, 2013-2015

(thousand m³)

	2013	2014	2015	Change (%) 2014-2015
Major importers				
China	16,910	17,462	17,638	1.0
Japan	7,245	5,989	5,770	-3.7
Egypt*	3,997	4,999	3,792	-24.1
Republic of Korea	1,520	1,724	1,882	9.2
Algeria	2,188	2,370	1,712	-27.8
Mexico*	1,130	1,134	1,353	19.3
Saudi Arabia	925	1,082	985	-9.0
United Arab Emirates	764	1,002	885*	-11.7
Major exporters				
Chile	3,111	3,596	3,139	-12.7
New Zealand	2,029	1,696	1,774	4.6
Brazil	716	958	1,266	32.2
Australia	241	363	297	-18.2

Note: *Estimate based on the sum of reported exports to the importing country.
Source: COMTRADE, 2016; Global Trade Atlas, 2016.

North African and Middle Eastern countries – particularly Algeria, Egypt, Saudi Arabia and the United Arab Emirates – continued to provide major markets for sawn softwoods in 2015, although import volumes declined in response to political instability and falling oil revenues. For example, sawn softwood imports fell by nearly 25% in 2015 in Egypt, the largest importer in the region, due primarily to political unrest.

Chile, New Zealand and Brazil (in descending order, by volume) were the only significant exporters of sawn softwoods outside the UNECE region in 2015. Chile's export markets are diversified, with significant volumes shipped to Asian, Latin American and Middle Eastern markets. Chile's exports to China – the single-largest national market – accounted for most of the decline in Chile's total sawn softwood exports in 2015, which occurred even though Chilean exporters increased their promotion of wood products in China and despite their ability to offer price-competitive products with shorter shipping times than European products (China Daily, 2015). New Zealand's major markets are more restricted and are predominantly in the Asia-Pacific region – Australia, China, the US and Viet Nam. Exports picked up in 2015 with some major reinvestments in sawmills, although there are concerns about the sustainability of New Zealand's radiata pine wood supply and the profitability of reinvestment in the plantation estate. Brazil's exports of sawn softwoods increased strongly in 2015, with shipments in the second quarter reaching their highest levels since 2006. Brazil's export volumes and prices have been affected in recent years by sharp fluctuations in the Brazilian currency (real) and by demand in the US, its major market. Exports have been encouraged by weak domestic consumption and uncertainty about the direction of the Brazilian economy, although the competitiveness of Brazilian companies is being reduced by record inflation, which is pushing up production costs.

Source: UNECE, 2015.

5.6 POLICY AND REGULATORY INFLUENCES ON THE SECTOR

The latest US–Canada Softwood Lumber Agreement[16] expired in mid-October 2015. No new agreement is expected in the one-year "stand-still period", during which there are no export duties on Canadian sawn softwood exports to the US. It is expected that, in mid-October 2016, the Coalition for Fair Lumber Imports will ask the US Department of Commerce to investigate Canadian lumber imports, which would result in the placing of preliminary countervailing and anti-dumping duties on Canadian lumber exports to the US in about March 2017. The last time this situation arose was in 2001, when the initial duty rates totalled 32%. The maximum duty rate (tied to sawnwood prices) under the previous nine-year agreement was 15% for companies in Alberta and British Columbia and 5% (and some quota volume restrictions) for the rest of Canada.

Efforts continue in North America to promote wood as a building material of choice. The US industry established the Softwood Lumber Board in 2011 for an initial five-year term. This is a mandatory promotion fund, or "check-off", authorized under the US Farm Bill, with the goals of increasing construction demand for sawn softwood, changing attitudes and buyer perceptions on wood, and converting projects from steel and concrete to wood. The Softwood Lumber Board operates with an annual budget of about $15 million funded by the industry, with a tariff of 35 cents per 1,000 board feet (approximately $0.22 per m^3, net size) levied on all suppliers to US markets, including importers, on volumes exceeding 15 million board feet (24,000 m^3) per supplier.

Note: The statistical annex of the ***Forest Products Annual Market Review 2015-2016*** is available at: www.unece.org/forests/fpamr2016-annex

[16] See section 2.2.2 of chapter 2, Policies shaping forest products markets, for more information on the Softwood Lumber Agreement.

5.7 REFERENCES

China Daily. 2015. Chile looks to bulk up wood-product exports to China. 06/04/2015. Available at: www.chinadaily.com.cn/world/2015livistsa/2015-04/06/content_20704809.htm

COMTRADE. 2016. United Nations Commodity Trade Statistics Database. Statistics Division. Available at: http://comtrade.un.org/db

EUWID Wood Products and Panels. 2010-2016. Various issues. Available at: www.euwid.com

Global Trade Atlas. 2016. Available at: www.worldtradestatistics.com/gta

IMF. 2016. World economic outlook: too slow for too long. April 2016. World Economic and Financial Surveys. International Monetary Fund (IMF). Available at: www.imf.org/external/pubs/ft/weo/2016/01/pdf/text.pdf

ITTO. 2016. Market Information Service. International Tropical Timber Organization (ITTO). Available at: www.itto.int/market_information_service

Japan Lumber Report. 2012-2016. Various issues. Available at: www.n-mokuzai.com

Random Lengths. 2016. Available at: www.randomlengths.com

Rosstat. 2016. Federal Service of State Statistics. Moscow. Available at: www.gks.ru

Statistics Canada. 2016. Available at: www.statcan.gc.ca/eng/start

UNECE/FAO. 2016. Timber database 2015-2016. Available at: www.unece.org/forests/fpm/onlinedata

US Department of Census. 2016. US housing starts. Available at: www.census.gov

WhatWood. 2016. Available at: www.whatwood.ru

Wood Markets China Bulletin. 2005-2016. Various issues. Available at: www.woodmarkets.com

Wood Markets Monthly. 2005-2016. Various issues. Available at: www.woodmarkets.com

World Resources Institute. 2016. Global timber and wood products market update. 16 March 2016.

6 SAWN HARDWOOD

Lead author: ***Rupert Oliver*** Contributing authors:

Fran Maplesden

HIGHLIGHTS

- Apparent consumption of sawn hardwood in the UNECE region was 35.6 million m^3 in 2015, a 0.9% rise compared with 2014 and the fourth consecutive year of increase.
- Falling consumption in Europe and the CIS in 2015 was offset by rising consumption in North America. European consumption of sawn hardwood decreased by 2.8%, to 12.2 million m^3, due mainly to a significant decline in Turkish consumption.
- Sawn hardwood production in the UNECE region increased by 1.8% in 2015, to 40.7 million m^3, with increases in all three subregions.
- EU furniture industry output was stagnant overall in 2015 – in large western European manufacturing countries it was 20-30% below levels seen before the global financial crisis.
- Multilayer parquet floors now account for about 84% of the European wood-flooring market by volume, while solid hardwood accounts for only 14%. There is a strong fashion trend for floorboards with grooves, knot holes and other irregularities.
- Sawn hardwood consumption in the CIS subregion fell by 25.9% in 2015, to 1.46 million m^3, following a 3.6% fall in 2014. Hardwood production increased in the subregion by 2.3%, to 3.37 million m^3, with exports taking up the slack.
- The weakness of the rouble encouraged a 50.5% increase in sawn hardwood exports by the Russian Federation in 2015, to 1.37 million m^3. China accounted for 1.17 million m^3 of these exports, 49% up from 2014 and by far the highest volume of this product ever shipped to China from the Russian Federation.
- North American sawn hardwood consumption increased by 5.7% in 2015, to 22.0 million m^3, with domestic sales in North America benefiting from rising new-home construction in the US.
- US sawn hardwood exports to countries outside North America decreased by 8.4% in 2015, to 3.0 million m^3, following five consecutive years of growth. Exports slowed in 2015 to all the leading markets, including (in descending order, by export volume) China (accounting for nearly half), Viet Nam, Mexico, the UK, Japan and Italy.
- Outside the UNECE region, China continued to dominate the sawn hardwood trade. China's imports of tropical and temperate sawn hardwoods were valued at $4.1 billion in 2015, marginally less than in 2014, when the value of imports had increased by 32% compared with 2013.
- Ukraine's export ban on unprocessed timber (published in 9 April 2015) includes logs, poles and sawnwood with thicknesses exceeding 70 mm and a moisture content above 22%.
- Innovations in the hardwood sector – such as new hardwood cross-laminated timber, glulam and laminated veneer lumber products – aim to extend the use of hardwood into new (notably structural) applications.

6.1 INTRODUCTION

Apparent consumption of sawn hardwood in the UNECE region was 35.6 million m³ in 2015, a 0.9% rise compared with 2014. Sawn hardwood consumption in the UNECE region has grown consistently each year since 2011. Falling consumption in Europe and the CIS in 2015 was offset by rising consumption in North America.

Sawn hardwood production increased in the UNECE region by 1.8% in 2015, to 40.7 million m³, with increases in all three subregions.

The recovery in sawn hardwood imports in the UNECE region, which began in 2014, continued in 2015, with imports increasing by 1.6%, to 6.58 million m³. Countries in the UNECE region exported 11.7 million m³ of sawn hardwood in 2015, up by 4.6% compared with 2014.

6.2 EUROPE

6.2.1 Consumption

European apparent consumption of sawn hardwood decreased by 2.8% in 2015, to 12.2 million m³ (table 6.2.1), continuing a slow decline in recent years. The downward trend in 2015 was due mainly to a significant decline in Turkish consumption, fed almost entirely by falling domestic production. Consumption in EU28 countries increased by 0.8% in 2015, to 9.5 million m³, benefiting from (albeit slow) growth in key sectors of the EU economy, including construction and furniture.

TABLE 6.2.1
Sawn hardwood balance, Europe, 2014-2016
(thousand m³)

	2014	*2015*	*2016f*	*Change (%) 2014-2015*
Production	13,159	12,984	13,190	-1.3
Imports	4,642	4,768	4,858	2.7
Exports	5,277	5,584	5,744	5.8
Apparent consumption	12,523	12,168	12,304	-2.8

Note: f = 2015 Committee on Forests and the Forest Industry forecast.
Source: UNECE/FAO, 2016.

Total construction output in the EU increased slightly in 2015 and is forecast to grow by 2-3% annually in the period 2016-2018. Much of the growth in European construction activity in recent years has been in repair, renovation and maintenance; these activities were responsible for roughly 60% of the total residential market in 2015 and have been key drivers of demand for sawn hardwood. Much of the growth in construction activity in 2016-2018, however, is likely to be in the residential new-build sector, driven partly by the large influx of migrants to western Europe (Euroconstruct, 2016).

The European furniture market, another key source of demand for sawn hardwood, has been recovering only slowly, growing by 1% in 2015 and unlikely to expand by more than 1.5% in 2016 (CSIL, 2016). The output of the furniture industry was stagnant in large western European manufacturing countries in 2015, at around 20-30% below the levels seen before the global financial crisis. In contrast, furniture industry output has been rising in parts of eastern Europe, notably Lithuania and Poland (AHEC, 2016).

The consumption of "real wood" flooring (not including laminate flooring) in the 17 countries covered by the European Federation of the Parquet Industry (FEP) increased by 0.5% in 2015, indicating stabilization in a market that had declined by 6% in the previous year. For the first time since the onset of the global financial crisis, southern European markets for hardwood flooring, particularly Spain, gained momentum in 2015 and the first half of 2016. Competition from other flooring types remains fierce, however, especially luxury vinyl tiles in Germany and ceramic tiles in southern Europe. It is becoming increasingly difficult for consumers to differentiate parquet from competitive flooring alternatives with a wood-look surface (FEP, 2016).

Source: AHEC, 2016.

There was no change in European hardwood fashion trends in 2015, which remain heavily oriented towards the "oak look". Oak continues to be used in over 70% of wood flooring manufactured in Europe; the share of tropical woods continues to decline and other temperate species account for only a small share (FEP, 2016). Multilayer parquet floors are increasingly dominant, accounting for around 84% (by volume) of the European wood-flooring market, while solid hardwood accounts for only 14%. There is a strong fashion for floorboards with grooves, knot holes and other irregularities and a trend for wood floors to be customized and exclusive (Interconnection Consulting, 2016).

6.2.2 Production and capacity change

European sawn hardwood production fell by 1.3% in 2015, to 13.0 million m³. The slight downward trend was due largely to a 17% decline in production in Turkey. EU28 production increased by 2.8%, to 10.3 million m³. Despite log shortages during the year, overall sawn hardwood output in Croatia, France and Germany was higher in 2015 than in 2014. Sawn hardwood production in Latvia – mainly of low-grade timber destined for pallets and other industrial applications – increased in 2015, while production was stable in Romania.

6.2.3 Prices

The strong fashion for oak, combined with the slow recovery of consuming sectors and the relative weakness of the euro against the US dollar (which has encouraged exports and increased prices for imported American alternatives) put pressure on supply and increased prices for European oak in 2015 and the first part of 2016. This was particularly true for thinner boards: due to high log prices, European oak mills curtailed the production of 1-inch boards, for which margins are lower despite shorter drying times, in favour of thicker boards (AHEC, 2016).

Prices for European sawn beech continued to strengthen in 2015, building on the gains made in 2014 after stagnation through most of 2013. European consumption of beech recovered slowly in 2015, particularly for superior colour grades of steamed beech, for which lead times lengthened (Brookes Brothers, 2015).

6.2.4 Trade

6.2.4.1 Imports

Total imports of sawn hardwood by European countries increased by 2.7% in 2015, to 4.79 million m^3. Due to the weakness of European currencies, however, the dollar value of imports fell by 8%, to $2.80 billion (although the euro value of trade increased by around 10%). Imports into several large western European consuming markets – including Belgium, France, Germany, the Netherlands and Spain – continued to recover in 2015. Imports also rose in large wood-manufacturing countries in eastern Europe, including Estonia, Lithuania, Poland and Slovenia. These gains offset a slight downturn in imports by Italy and the UK.

Exchange-rate volatility was a major determinant of the volume and direction of sawn hardwood trade by European countries in 2015 and the first half of 2016. The value of the euro fell by 20% against the US dollar between July 2014 and March 2015 and remained at the lower level throughout 2015 and into the first quarter of 2016. In the same period, however, the euro strengthened by 50% against the Ukrainian hryvnia and by 40% against the rouble. The UK pound followed a different trajectory, weakening against the dollar in 2015 and strengthening against the euro before falling sharply against both currencies after the Brexit vote on 23 June 2016.

Source: AHEC, 2016.

Overall, these changes favoured trade in European hardwoods at the expense of American hardwoods and particularly strengthened European imports from CIS countries. In 2015 there was rising trade in ash and oak from Ukraine (destined mainly for Italy and Poland) and in aspen and birch from Belarus and the Russian Federation (destined mainly for Estonia, Germany and Lithuania). US sawn hardwood exports to Europe fell by 11.1% in 2015, to 348,100 m^3. Coupled with rising dollar prices for American hardwoods, the weak euro meant that some species and grades were up to 40% more expensive for buyers in the eurozone in 2015 than in 2014.

Europe's imports of tropical hardwoods, now sourced mainly from African countries in which prices are quoted in euros, were less affected by currency volatility. Although still well below the volumes of a decade before, EU tropical sawn hardwood imports recovered ground, rising by 5% in 2014, to 977,000 m^3, and by 10.5% in 2015, to 1.08 million m^3.

6.2.4.2 Exports

Following a 17% increase in 2014, sawn hardwood exports by European countries increased by 5.8% in 2015, to 5.58 million m^3, assisted by currency movements and rising consumption in key export markets. Exports by Croatia, the leading exporter among European countries, continued to increase in 2015, rising by 4.4%, to 907,000 m^3, driven mainly by a recovery in shipments to Egypt. Exports of sawn hardwood from Romania increased by 7%, to 806,000 m^3. That country's exports to Egypt, its largest market, increased sharply in 2015, offsetting a slight decline in exports to China. Exports by Germany decreased by 2.7% in 2015, to 691,000 m^3, with rising sales to Poland and North America insufficient to offset declining sales to China, the Netherlands, the UK and Viet Nam.

6.3 COMMONWEALTH OF INDEPENDENT STATES

Sawn hardwood consumption in the CIS subregion fell by 25.9% in 2015, to 1.46 million m^3, following a 3.6% fall in the previous year. Hardwood production increased by 2.3% in 2015, to 3.37 million m³. A sharp decline in economic activity in the CIS subregion, driven partly by falling oil prices and political instability in Ukraine, combined with currency weaknesses, led to the export of a much larger share of sawn hardwood production: exports from the subregion increased by 41.6% in 2015, to 2.0 million m³. Sawn hardwood imports by CIS countries also increased in 2015 (by 1.2%) but remain negligible, at 90,570 m³ (table 6.3.1).

TABLE 6.3.1
Sawn hardwood balance, CIS, 2014-2016
(thousand m³)

	2014	2015	2016f	Change (%) 2014-2015
Production	3,298	3,374	3,374	2.3
Imports	89	91	99	2.2
Exports	1,413	2,000	2,028	41.5
Apparent consumption	1,975	1,464	1,444	-25.9

Note: f = 2015 Committee on Forests and the Forest Industry forecast.
Source: UNECE/FAO, 2016.

Apparent consumption of sawn hardwood in the Russian Federation fell by 29% in 2015, to 1.14 million m³. Economic conditions in the Russian Federation deteriorated, with a 3.7% decline in GDP – the biggest contraction in six years – due mainly to declining oil prices, a 15.3% decline in retail sales (the biggest fall in more than two decades) and a 4.5% fall in industrial production (Financial Times, 2016). However, the weakness of the rouble encouraged a 50.5% increase in sawn hardwood exports by the Russian Federation in 2015, to 1.37 million m³. Exports to China were 1.17 million m³, an increase of 49% over 2014 and by far the largest quantity of Russian sawn hardwood ever shipped to China. Exports also increased to Estonia, Kazakhstan, Latvia, Lithuania and Poland (Global Trade Atlas, 2016).

Economic problems in Ukraine, which experienced a 9.9% fall in real GDP in 2015, led to a 37% fall in sawn hardwood consumption in that country in 2015. The problems continued in 2016, particularly with the escalation of a trade dispute with the Russian Federation following Ukraine's free-trade deal with the EU (introduced at the start of the year). The weak hryvnia, combined with measures to restrict log exports from Ukraine, however, contributed to a 23.8% increase in exports, to 437,000 m³ in 2015. Ukraine's sawn hardwood exports, now strongly oriented towards EU countries, increased in all the main markets, including Germany, Italy, Lithuania, Poland and Romania.

6.4 NORTH AMERICA

6.4.1 Consumption

North American sawn hardwood consumption increased by 5.7% in 2015, to 22.0 million m³ (Table 6.4.1). The North American market continued to benefit from rising new-home construction in the US in 2015, although there were signs of slowing consumption in some sectors. Sawn hardwood production increased in 2015, but at a slower pace than in the previous two years. Exports were hit in 2015 by the slowdown in the Chinese economy and by the strong US dollar. Total imports, including the large cross-border trade between Canada and the US, were stable in 2015, but imports from outside the subregion continued to rise (table 6.4.1).

Sawn hardwood consumption in the US increased by 5.1% in 2015, to 20.1 million m³; it increased in the furniture, millwork and railway ties subsectors, although these gains were partly offset by declines in the consumption of hardwood pallets, board roads, cabinets and flooring (Hardwood Market Report, 2016) (graph 6.4.1).

Source: AHEC, 2016.

US housing starts continued to rise at double-digit rates in 2015, boosting demand for millwork and furniture. The pace of growth in sawn hardwood consumption in these subsectors has been slower than the rate of construction growth overall, however, due to competition from imports and other materials. The potential for the return of large-scale wood furniture manufacturing in the US, widely forecast as a response to consumer demand for quality bespoke products and services, remains largely unrealized. Imported furniture, particularly from China and Viet Nam, continues to dominate. US sawn hardwood also lost market share to imported and engineered products in the cabinets and flooring subsectors in 2015.

TABLE 6.4.1

Sawn hardwood balance, North America, 2014-2016

(thousand m³)

	2014	*2015*	*2016f*	*Change (%) 2014-2015*
Production	23,490	24,323	24,256	3.5
Imports	1,741	1,718	1,702	-1.3
Exports	4,463	4,086	3,728	-8.4
Apparent consumption	20,768	21,955	22,231	5.7

Note: f = 2015 Committee on Forests and the Forest Industry forecast.
Source: UNECE/FAO, 2016.

Demand for US sawn hardwood in the wooden pallet segment came under intense competitive pressure from alternative materials, notably pine, in 2015. Demand also slowed in the board roads segment due to reduced oil and gas field construction (graph 6.4.1).

With the exception of the hardwood pallet segment, consumption trends in the US improved in the first half of 2016. Steady gains are expected in the second half of the year based on continuing improvement in residential construction (Hardwood Market Report, 2016).

GRAPH 6.4.1

US sawn hardwood consumption, by segment, 2007-2015

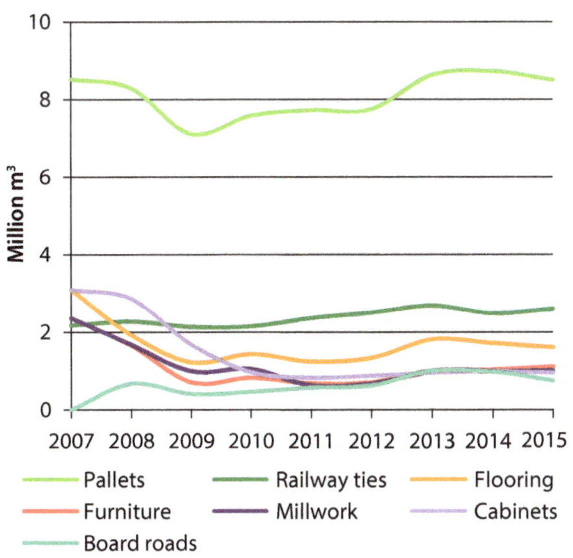

Source: Hardwood Market Report, 2016.

Sawn hardwood consumption increased in Canada in 2015, boosted by a continuing rise in residential construction in North America and by low interest rates.

6.4.2 Production and capacity change

US sawn hardwood production has increased consistently in recent years; in 2015 it was at 22.6 million m³, 27.6% higher than in 2011. The annual rate of growth slowed to 2.4% in 2015, however, compared with average growth of 8% in the previous three years. Growth was curtailed in 2015 in response to a downturn in domestic demand for pallets and board roads. Residential solid-wood-flooring manufacturers also overbought early in the year and suffered from overstocking. At the same time, export demand for higher-grade sawnwood weakened with the slowdown of the Chinese economy and the stronger US dollar.

6.4.3 Prices

Prices for kiln-dried US hardwood, which increased sharply between 2013 and 2014, weakened by around 20% in 2015 in response to improved supply and slowing demand in China. A larger downturn in prices was forestalled by a slowdown in production in the second and third quarters of the year. Prices had stabilized by the beginning of 2016 as both domestic and export demand began to recover (graph 6.4.2) (Weekly Hardwood Review, 2016).

GRAPH 6.4.2

Prices for selected hardwood species in the US, 2009-2016

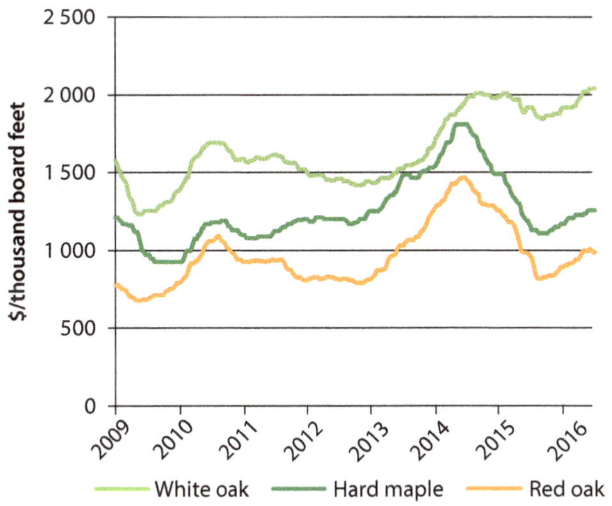

Note: Nominal prices. Data until 24 June 2016.
Source: Weekly Hardwood Review, 2016.

6.4.4 Trade

6.4.4.1 Imports

The significant cross-border trade in sawn hardwood between Canada and the US, which rose between 2011 and 2014, dipped in 2015. The US imported 363,000 m³ of sawn hardwood from Canada in 2015, down by 11% compared with the previous year. Canada imported 525,000 m³ of sawn hardwood from the US in 2015, 16% down from 2014 (Global Trade Atlas, 2016).

After rising by 18% in 2014, US imports of temperate sawn hardwood from outside North America decreased by 24% in 2015, to 144,128 m³, driven by a significant fall in imports from Uruguay (mainly Eucalyptus grandis) and Italy. Imports from China and Germany, the leading suppliers to the US from outside the subregion, were stable in 2015. The US imports

300,000-400,000 m³ of tropical sawn hardwood per year.[17] Tropical hardwoods consist mainly of decking and flooring species from Brazil, Cameroon and Malaysia, and balsa from Ecuador.

Canadian imports of sawn hardwood from outside the subregion decreased by 30% in 2015, to 53,000 m³, and contributed only a small share of total consumption. Most of the decrease was due to a sharp fall in imports from Ecuador, primarily balsa. Imports from Brazil, the leading supplier to Canada from outside the subregion, were stable in 2015 (Global Trade Atlas, 2016).

6.4.4.2 Exports

US sawn hardwood exports to countries outside North America decreased by 8.4% in 2015, to 3.0 million m³, following five consecutive years of growth that culminated in a 19% increase in 2014. Exports slowed to all the leading markets, including (in descending order, by export volume) China, Viet Nam, Mexico, the UK, Japan and Italy. Indonesia and Spain were the only large US export markets recording increases in 2015. Despite the overall decline, US exports in 2015 were still the second highest ever recorded.

In volume terms, China accounted for 47% of US sawn hardwood exports in 2015, Canada for 15%, Southeast Asia for 13%, Europe for 10% and Mexico for 8%. Red oak was the leading export species in 2015, accounting for 25% by volume, followed by white oak (16%), tulipwood (15%) and ash (10%). The share of red oak and white oak in exports increased in 2015, while the share of tulipwood declined (USDA, 2016).

US sawn hardwood exports were up by 8% in the first four months of 2016, year-on-year. Exports increased to China, Germany, Mexico, the UK and Viet Nam but continued to decline to Italy and Japan (Global Trade Atlas, 2016).

Canadian producers focused heavily on North American markets in 2015. Canadian exports of sawn hardwood to countries outside the subregion increased by only 3%, to 162,000 m³. Canada exported 80,000 m³ to China/Hong Kong SAR in 2015, 9% higher than in 2014, and 29,000 m³ to the EU, down by 12%. Canada's total sawn hardwood exports were up by 5% in the first five months of 2016 compared with the same period in 2015 (Global Trade Atlas, 2016).

[17] The actual volume varies within these bounds and is uncertain due to irregular large volumes reported in US official data under Harmonized Trade Schedule code 4407.99.01.93 as imported from tropical countries as "other nonconiferous" sawnwood. Unit values of imports of sawn hardwood from tropical countries reported under this code are frequently very low and the volumes are not mirrored in the sawn hardwood export data of partner tropical countries. Correcting for irregular data under 4407.99.01.93, it is estimated that US tropical sawn hardwood imports increased by 3.6% in 2015, from 333,400 m³ in 2014 to 345,500 m³ in 2015.

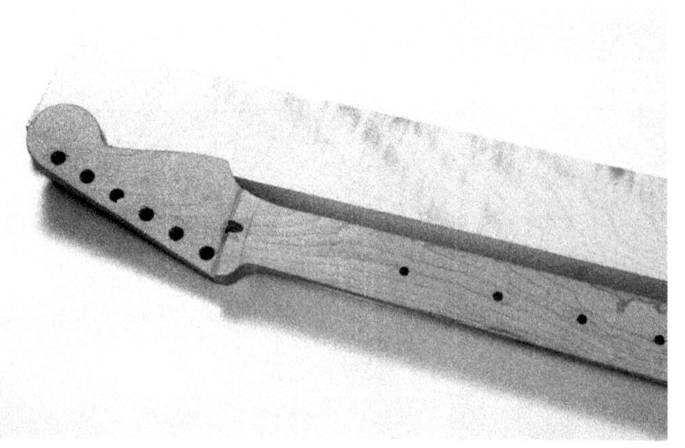

Source: AHEC, 2016.

6.5 EXTRAREGIONAL INFLUENCES AFFECTING THE UNECE REGION

Outside the UNECE region, China continued to dominate the sawn hardwood trade, influencing the direction of trade of both hardwood logs and sawnwood. China's imports of tropical and temperate sawn hardwoods were valued at $4.1 billion in 2015, marginally less than in 2014, when imports increased by 32% by value compared with 2013. Demand growth began to slow at the end of 2014, but demand for sawn hardwood was less affected by slowing construction activity in 2015 than by demand for sawn softwood, for which the value of imports dropped by 11% in 2015 (Global Trade Atlas, 2016).

The world tropical sawn hardwood trade continued to focus on China, the dominant importer, in 2015 and, to a lesser extent, on other markets in the Asia-Pacific region – Singapore, Taiwan Province of China, Thailand and Viet Nam. Malaysia and Thailand were the major tropical hardwood exporters in 2015.

China's major supplier of tropical sawn hardwood in 2015 was Thailand (64%), with significant volumes also imported (in descending order, by volume) from Gabon, Indonesia, Malaysia, the Philippines, Viet Nam, the Lao People's Democratic Republic and Cameroon. China's imports from Gabon and Thailand increased considerably in 2015, by 41% and 37% respectively, while supplies declined from most of the other main tropical suppliers (the exception being Cameroon). Imports from Thailand are predominantly of lower-value rubberwood, but Africa's supplies are mainly of high-value specialty timbers for the high-end market, with prices remaining relatively stable in the year to March 2016 (ITTO, 2016).

African tropical sawn hardwood exports, which previously went mainly to European markets, have shifted to China. There was a minor increase in exports of African sawn hardwoods to European and US markets in 2015, but not to previous levels. Thailand's exports, which totalled over 3 million m³ in 2015, go overwhelmingly (99%) to China; Malaysia exported to a greater range of markets, with EU countries (particularly Belgium,

France, the Netherlands and the UK) among other important destinations.

6.6 POLICY AND REGULATORY INFLUENCES

Ukraine published a law on 9 April 2015 imposing an export ban on unprocessed timber for ten years, effective from 1 November 2015 for all species except pine (the export of which will be banned from 1 January 2017). In addition to logs and poles, the ban extends to sawnwood with thicknesses exceeding 70 mm and moisture content greater than 22%. Ukraine also added oak to its list of "rare and valuable" timber species, for which controls are imposed on a wider range of secondary and tertiary processed products. Although these laws only became effective later in 2015, their publication had an immediate impact, encouraging greater exports of sawn hardwood at the expense of logs. The EU imported only 29,000 m^3 of Ukrainian oak logs in 2015, down from 82,000 m^3 in the previous year and 200,000 m^3 before the global financial crisis.

Source: UNECE, 2016.

Plant health issues are becoming increasingly prominent in the international hardwood trade. Trade in American ash has been particularly affected by restrictions designed to control the spread of the emerald ash borer (EAB). In January 2016, new rules came into force for the treatment of ash imported into the EU from North America, with zero tolerance for residual bark and wane on any ash wood sourced from an area not listed as EAB-free in EU legislation. The rules have caused a significant decrease in the availability of American ash to EU importers; US exporters are charging more for supplying fully square-edged products due to the extra sorting and reduced yield (AHEC, 2016).

Laws such as the EU Timber Regulation, which has been in force since March 2013, and the US Lacey Act amendment of May 2008 have heightened the sensitivity of the sawn hardwood sector to illegal harvesting and encouraged measures to demonstrate a negligible risk that wood has been obtained from illegal sources. The overall impact on the trade of timber harvested in the UNECE region has been minimal to date; it has been more significant for tropical hardwoods, for example by focusing procurement on a narrower range of tropical suppliers who can provide credible assurances of legal origin. Various high-profile cases, however, indicate a non-negligible risk of illegal harvesting in some UNECE countries and that sanctions for non-compliance may be serious (World Resources Institute, 2015). In 2015, there were signs that regulatory authorities in Europe and North America were ramping up enforcement activities (Forest Trends, 2016). The laws are expected to have an increasing impact on trade.

6.7 INNOVATION IN THE SECTOR

Innovations in the sawn hardwood sector aim to extend uses into new applications, notably structural applications through the development of new products made of hardwood cross-laminated timber, glulam and laminated veneer lumber. Work continues in both Europe and North America to broaden the use of temperate hardwoods in external environments through thermal and (various forms of) chemical modification. Another innovation has been to increase the efficiency of wood processing. The hardwood company Danzer, for example, has developed machines capable of slicing 3-mm-thick wooden lamellas for engineered flooring products, with almost no loss of raw material. These machines replace traditional saws, which waste up to 50% of wood material as sawdust. Danzer has also pioneered "curve-saw" technology that can produce flat boards from swept hardwood logs (which occur often in nature), thereby increasing yield (Danzer, 2016).

Note: The statistical annex of the *Forest Products Annual Market Review 2015-2016* is available at: www.unece.org/forests/fpamr2016-annex

6.8 REFERENCES

AHEC. 2016. Personal communications with D. Venables and M. Snow, American Hardwood Export Council (AHEC), June 2016.

Brookes Brothers. 2015. General hardwood market conditions. Company report by D. McCarthy, Group Purchasing Director. October 2015.

CSIL. 2016. Outlook for the global market in 2016. Article for www.worldfurnitureonline.com, March 2016. Center for Industrial Studies (CSIL).

Danzer. 2016. 15 million euro investment program: Danzer invests further in deck layer production for engineered flooring. Press release, 1 January 2015.

Euroconstruct. 2016. Ongoing recovery in European construction. Press release on the occasion of the 81st Euroconstruct Conference in Dublin, Ireland, 10 June 2016.

FEP. 2016. Facing the future with optimism: the European parquet industries in 2015. Press release, 23 June 2016. European Federation of the Parquet Industry (FEP).

Financial Times. 2016. Russian GDP contracted 3.7% in 2015. FastFT article, 25 January 2016.

Forest Trends. 2016. Survey shows progress on US, EU, and Australian enforcement actions against illegal logging – what role can China play? Blog post by K. Canby, J. Saunders and N. Basik Treanor, 21 June 2016.

Global Trade Atlas. 2016. Available (to subscribers) at: www.gtis.com/gta

Hardwood Market Report. 2016. Available (to subscribers) at: www.hmr.com

Interconnection Consulting. 2016. Authenticity brought by wooden floors. Press release, 7 March 2016.

ITTO. 2016. Market Information Service. International Tropical Timber Organization (ITTO). Available at: www.itto.int/market_information_service

UNECE/FAO. 2016. TIMBER database. Available at: www.unece.org/forests/fpamr2013-annex

USDA. 2016. Global Agricultural Trade System. US Department of Agriculture (USDA). Available at: http://apps.fas.usda.gov/gats/default.aspx

Weekly Hardwood Review. 2016. Available (to subscribers) at: www.hardwoodreview.com

World Resources Institute. 2015. $13 million fine for lumber liquidators shows U.S. Lacey Act's clout. October 2015. Available at: www.wri.org/blog/2015/10/13-million-fine-lumber-liquidators-shows-us-lacey-acts-clout

7 WOOD-BASED PANELS

Lead authors: ***Ivan Eastin***

Contributing authors:

Marion Le Roy,
Fran Maplesden
and Igor Novoselov

HIGHLIGHTS

- The market for wood-based panels in Europe was mixed in 2015, with consumption increasing for plywood and oriented strandboard and declining for particle board and medium-density fibreboard. Near-term expectations are cautiously optimistic, except for medium-density fibreboard.
- The construction sector in Europe turned positive in 2015 after years of contraction. The outlook is positive, with residential building activity projected to rise.
- 2015 was a successful year for export-oriented wood-based panel producers in the Russian Federation, where declining global prices were offset by a weak rouble and slightly improved domestic sales.
- The heavy devaluation of the rouble strongly reduced Russian imports of wood-based panels from Europe and North America, to the benefit of domestic manufacturers.
- Weak demand for real estate in the Russian Federation has been influenced by both a decline in real incomes and high mortgage rates.
- Wood-based panel consumption increased in North America in 2015, with the exception of particle board, which was down slightly.
- Capacity utilization in North America declined for plywood in 2015, to 74% (from 76% in 2014), but was steady for oriented strandboard, at 71%.
- Demographic shifts in North America are driving a shift from single-family to multifamily homes. Moreover, the average size of single-family homes is likely to shrink in the future. Both trends would likely reduce demand for panels in the future.
- Prices for non-structural panels in North America were generally flat through 2015. Among structural panels, there was a strong price increase for oriented strandboard, whereas plywood prices continued their decline from mid-2014.
- The supply of wood residuals in North America tightened in 2015 and prices increased, with demand – most notably for wood pellets in the UK – rising in global bioenergy markets.

7.1 INTRODUCTION

The wood-based panels market was remarkably similar in 2014 and 2015, with the exception of Europe, where trends in production, imports and consumption were down in 2015. The single bright spot in Europe was the OSB subsector, in which production grew by 2.0%. European production in the other product subsectors was unremarkable: plywood was almost unchanged (+0.3%), fibreboard increased slightly (+0.9%), and particle board declined (-0.7%).

The big news in the CIS subregion continued to be the weakness of the rouble, which depressed imports substantially (down by 15.6% in 2015) while fuelling a large expansion of exports (up by 16.5%). There were substantial production increases in both the OSB (+120%) and MDF (+26.1%) subsectors in the CIS in 2015 as new plants continued to open and existing plants expanded.

The market for wood-based panels in North America moderated slightly in 2015, with overall wood-panel consumption increasing by 4.2% (compared with a 5% increase in 2014). Weak production trends in the plywood (0.8%) and particle board (+0.2%) subsectors were offset by gains in the production of fibreboard (+1.3%) and OSB (+2.4%).

The various fibreboard products (e.g. hardboard, MDF, high-density fibreboard and insulating board) are easily misclassified; thus, this chapter presents general trends for fibreboard, although some subproducts are mentioned specifically.

Source: UNECE/FAO, 2016.

7.2 EUROPE

The European economy benefited from several positive factors in 2015, including low oil prices, a favourable euro exchange rate, supportive monetary policy measures and, in some countries, increased public expenditure associated with the inflow of asylum seekers. Economic indicators continue to move encouragingly in Europe in 2016 but at a slower pace, resulting in more moderate growth. There were only minor changes in the wood-based panels market in Europe in 2015 compared with 2014. Apparent consumption increased overall by 0.7% (table 7.2.1); it decreased slightly for particle board (0.6%) and MDF (-2.6%) and increased marginally for plywood (+0.9%) and significantly for OSB (+5.5%).

TABLE 7.2.1

Wood-based panel balance, Europe, 2014-2016

(thousand m³)

	2014	*2015*	*2016f*	*Change (%) 2014-2015*
Production	68,945	69,014	69,164	0.1
Imports	32,173	32,651	32,858	1.5
Exports	33,175	33,272	35,614	0.3
Apparent consumption	67,942	68,392	66,408	0.7

Note: f = 2015 Committee on Forests and Forest Industry forecast.
Source: UNECE/FAO, 2016.

Due to methodological and other differences, information from the private sector indicates a more positive view of trends in wood-based panels in Europe than that obtained from the official statistics presented in this section (European Panel Federation, 2016).

7.2.1 Consumption

Particle board. Apparent particle board consumption decreased by 0.6% in Europe in 2015, to 35.2 million m^3, down from 35.4 million m^3 in 2014. The top five consuming markets for particle board were (in descending order): Germany, Poland, Turkey, Italy and the UK, together accounting for 60% of European consumption (UNECE, 2016). The furniture industry remained the largest end-user of particle board in Europe in 2015; other applications were in the building industry (including doors and flooring applications) and packaging (European Panel Federation, 2016).[18] Particle board consumption is expected to remain stable in 2016 (UNECE, 2016).

Fibreboard. Apparent fibreboard consumption decreased by 1.1% in Europe in 2015, to 19.4 million m^3 (UNECE, 2016). Turkey was by far the largest consumer of fibreboard in the subregion, consuming 4.6 million m^3 in 2015, a decrease of 3.8% over 2014. The second-highest consumer was Germany, with a consumption of 2.9 million m^3 (down by 6.5%), followed by Poland, at 2.9 million m^3 (down by 2%), and the UK, at 1.4 million m^3 (down by 3%). Apparent consumption of fibreboard increased by 8% in Italy, to 1.3 million m^3.

MDF. Apparent consumption decreased by 2.6% in Europe in 2015, to 14.1 million m^3, and is expected to decrease again (by 5.4%) in 2016 (UNECE/FAO, 2016). The furniture (45%) and laminate flooring (34%) subsectors were the main users of MDF panels in 2015. Despite the popularity of renovation and

[18] The European Panel Federation reports information on 27 European countries: Austria, Belgium, Bulgaria, Croatia, the Czech Republic, Denmark, Estonia, Finland, France, Germany, Greece, Hungary, Ireland, Italy, Latvia, Lithuania, Luxembourg, Norway, Poland, Portugal, Romania, Slovakia, Slovenia, Spain, Sweden, Switzerland and the UK.

do-it-yourself markets, sales to the building sector amounted to only 18%. The remaining 3% of European MDF production was consumed in the production of mouldings, panelling and small cabinets for home entertainment systems, among other products (European Panel Federation, 2016).

Hardboard and insulating board. Apparent consumption of both hardboard and insulating board increased in 2015, by 0.6% (to 2.5 million m^3) and 4.8% (to 3 million m^3), respectively. The main end uses for hardboard were packaging and do-it-yourself products (32% and 20%, respectively). Furniture products accounted for 19% of consumption, construction applications for 9% and automotive products for 5%. Softboard (classified under insulation board) was consumed mainly as rigid building shells (47%), flexible building shells (31%), rigid underlays (10%) and standard boards (7%), with other specialty softboards accounting for the remainder (European Panel Federation, 2016).

OSB. After several years of contraction, building activity increased for a second consecutive year in 2015, although growth was stronger in some eastern European countries than in western Europe. Overall European consumption of OSB increased by 5.5% in 2015, to almost 5 million m^3 (UNECE/FAO, 2016). Germany remained the dominant market, consuming 1.4 million m^3 in 2015 (up by 14.3% from 2014 and representing 28% of total European consumption). The UK (517,000 m^3), Poland (486,000 m^3), France (420,000 m^3) and Romania (382,000 m^3) were the next-largest markets for OSB in Europe. European consumption of OSB is expected to grow by 2.6% in 2016.

Plywood. European consumption of plywood increased by 0.9% in 2015, to more than 8 million m^3 (UNECE, 2016). The top five consuming nations were the UK (1.4 million m^3, up by a strong 7.1%), Germany (1.2 million m^3), Romania (596,000 m^3), France (506,000 m^3) and Italy (493,000 m^3), together accounting for more than half the plywood consumed in the subregion (UNECE, 2016). The main plywood applications in Europe in 2015 were construction (40%), furniture (28%), transport (14%) and packaging (9%) (European Panel Federation, 2016). European plywood consumption is expected to increase by 1% in 2016 (UNECE/FAO, 2016).

An estimated 5.0 million m^3 of non-coniferous plywood was consumed in Europe in 2015, an increase of 1.2% over 2014. The consumption of coniferous plywood was 3.1 million m^3, up by 0.5% compared with 2014.

7.2.2 Production and capacity utilization

Graph 7.2.1 shows that particle board comprised more than half the total wood-based panel production in Europe in 2015, with fibreboard accounting for one third and OSB for a little over 8%.

GRAPH 7.2.1

Wood-based panel production in Europe, 2015

(million m^3)

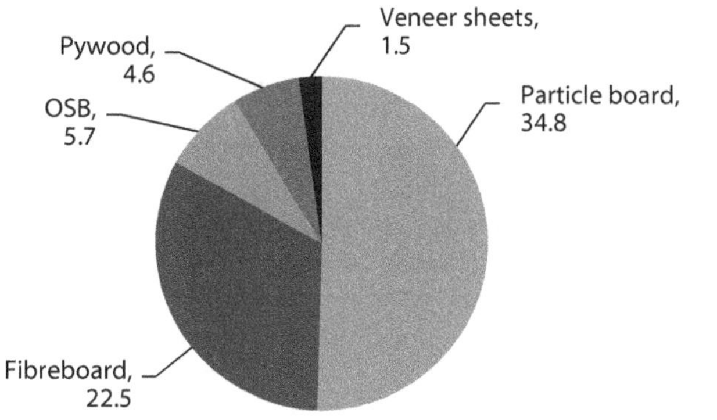

Notes: Total wood-based panel production in Europe in 2015 = 69.0 million m^3. The fibreboard component comprises MDF (87%), hardboard (6%) and insulating board (7%).
Source: UNECE/FAO, 2016.

Particle board. European particle board production decreased slightly (by 0.7%) in 2015, to 34.8 million m^3, still well below the output peak of 37.8 million m^3 achieved in 2007. Significant production increases were recorded in Bosnia and Herzegovina (+53.8%), Estonia (+37.3%) and Italy (+3.5%). In contrast, production declined substantially in Slovakia (-16.7%), Slovenia (-13.3%) and Germany (-2.4%).

Despite the decline in production, Germany remained Europe's largest particle board producer in 2015, at 5.5 million m^3, accounting for 15% of total European production. Poland (4.39 million m^3) and Turkey (4.36 million m^3) also produced significant volumes of particle board, followed by France and Romania; together, these five countries accounted for 56% of total European production in 2015. Particle board production is expected to increase by about 5% in 2016, to 36.7 million m^3 (UNECE/FAO, 2016).

Total particle board production capacity in the EPF member countries (de facto EU28 plus European Free Trade Association – EFTA – countries) decreased by 3.9% (1.5 million m^3) in 2015, to 37.2 million m^3. European particle board production capacity is expected to increase marginally (by 1%) in 2016 (European Panel Federation, 2016).

Fibreboard. Fibreboard production grew by 0.9% in Europe in 2015, to 22.5 million m^3. With a stable output of about 5.3 million m^3, Germany remained the largest European fibreboard producer in 2015. The other main producers were Turkey, followed by Poland; the combined output of these three countries accounted for 63% of total European fibreboard production in 2015 (UNECE/FAO, 2016).

European production of MDF was stable (down by only 0.2%) in 2015, at 16.3 million m^3, and is expected to remain so in 2016. Turkey and Poland accounted for almost half Europe's MDF production in 2015; Turkish production decreased by 2.2% and Polish production increased by 2.7%. The overall MDF capacity utilization rate in the EU28 plus EFTA countries increased from 77% in 2014 to 79% in 2015 (European Panel Federation, 2016).

European hardboard production increased by 0.9% in 2015, to 3.0 million m^3, with Germany accounting for 77%. European production of insulating board increased in 2015, for the fourth year in a row; it was up by 1.5% in 2015, to just over 3 million m^3.

Rigid softboard accounted for 65% of the softboard output in the EU28 plus EFTA countries in 2015, and flex softboard accounted for 35%. Installed production capacity decreased slightly for rigid softboard in 2015, to 3.5 million m^3, and also for flex softboard, to 1.85 million m^3. Germany and Poland are the biggest European producers of both types of softboard (European Panel Federation, 2016).

OSB. OSB production increased by 2.0% in 2015, to 5.7 million m^3. Germany and Romania have the largest OSB production capacities in the subregion (UNECE, 2016). Due primarily to an expansion of capacity in Poland, Europe's OSB production capacity increased slightly in 2015. If investments in Belgium, Hungary and Ireland are confirmed and implemented, production capacity in the EU28 plus EFTA countries could exceed 6.5 million m^3 in 2017 in the wake of additional investments in the UK. Given that new projects have also been announced in Belarus and possibly Turkey, there could be a capacity surplus in the near-to-medium term (European Panel Federation, 2016). OSB production is expected to increase by 1.3% in 2016 (UNECE/FAO, 2016).

Plywood. Plywood production was stable in Europe in 2015, at 4.6 million m^3, but it is expected to increase by 1% in 2016 (UNECE/FAO, 2016). The majority (59%) of European plywood was produced from temperate hardwoods in 2015, consisting primarily of birch, poplar and beech. Coniferous plywood – spruce and pine – represented 33% of total plywood production, while tropical hardwoods (e.g. okoumé, ilomba and ceiba) accounted for 8% (European Panel Federation, 2016). The top five plywood-producing countries in 2015 were Finland (1.15 million m^3), Romania (670,000 m^3), Poland (419,000 m^3), Spain (284,000 m^3) and Latvia (250,000 m^3).

Source: UNECE/FAO, 2016.

7.2.3 Trade

7.2.3.1 Imports

Particle board. Europe was a net exporter of particle board in 2015 (with net exports of 1.3 million m^3). Imports decreased by 0.2%, to 11.1 million m^3.

Despite a decrease in its imports in 2015, Germany remained the largest European importer of particle board, at 2.2 million m^3, followed by Poland and Italy. European imports of particle board are expected to decrease slightly (0.6%) in 2016.

Fibreboard. European imports of fibreboard increased slightly (by 0.6%) in 2015, to 9.8 million m^3. MDF imports were down by 1.6%, to 6 million m^3, but imports of hardboard and insulating board both increased, by 0.6% (to 2.5 million m^3) and 5.2% (to 1.6 million m^3), respectively. Imports of MDF to Europe are expected to decrease by 1.3% in 2016 (UNECE/FAO, 2016).

OSB. Imports of OSB into Europe increased by 4.9% in 2015, to 3 million m^3; the top importing country was Germany, with 637,000 m^3. OSB imports to Europe are expected to increase by 1.8% in 2016 (UNECE/FAO, 2016).

Plywood. Europe imported 7.5 million m^3 of plywood in 2015, up by 2.2% from 2014 and led by the UK (1.5 million m^3) and Germany (1.4 million m^3). Of the total, coniferous plywood imports accounted for 3 million m^3 (up by 0.7%), temperate hardwood plywood for 4.5 million m^3 (up by 3.2%), and tropical hardwood plywood for 991,000 m^3 (up by 4.4%). Imports of plywood to Europe are expected to increase by 2.4% in 2016 (UNECE/FAO, 2016).

7.2.3.2 Exports

Particle board. European particle board exports have been stable in recent years; they increased by just 0.4% in 2015, to 12.4 million m^3.

Austria remained the largest exporter of particle board in 2015, at 1.83 million m^3 (up by 1.3%), followed by Germany and France (both at 1.75 million m^3, with Germany's volume up by 1.3%), and Romania, at 1.28 million m^3. Together, these four countries produced more than half the particle board exported from European countries in 2015. European particle board exports are expected to increase only slightly (by 0.3%) in 2016 (UNECE/FAO, 2016).

Fibreboard. European exports of fibreboard increased by 2.6% in 2015, to 12.7 million m^3. MDF exports increased by 3.3%, to 8.1 million m^3, hardboard exports increased by 3.1%, to 2.9 million m^3, and exports of insulating board decreased by 1.6%, to 1.7 million m^3. European MDF exports are expected to increase by 8.3% in 2016, to 8.8 million m^3 (UNECE/FAO, 2016).

OSB. European OSB exports fell by 0.9% in 2015, to 3.7 million m^3. European OSB exports are expected to remain stable in 2016.

Source: APA, 2016.

Plywood. European countries exported 4.1 million m^3 of plywood in 2015, up by 2.2% from 2014. Of the total, coniferous plywood accounted for 2 million m^3 (up by 2%), non-coniferous for 2.1 million m^3 (up by 2.4%), and tropical hardwood plywood for 412,000 m^3 (up by 2.9%). The top plywood exporters in 2015 were Finland (981,000 m^3), Belgium (369,000 m^3) and Germany (332,000 m^3). Europe's plywood exports are expected to increase by 11% in 2016, to 4.5 million m^3 (UNECE/FAO, 2016).

7.3 COMMONWEALTH OF INDEPENDENT STATES, WITH A FOCUS ON THE RUSSIAN FEDERATION

The Russian wood-based panels industry was relatively successful in 2015 due to a weak rouble, which helped propel exports while limiting the competitiveness of imported panel products. For the last four years, revenue from sales of wood-based panels has grown by an average of 17% per year, due largely to inflation in the domestic market and the devaluation of the rouble.

Investment remained strong in the Russian wood-based panels sector, despite challenging economic conditions. For example, the value of investments in the plywood sector in 2015 was 6.3 billion roubles, and investment worth 18 billion roubles (almost double that of 2014) was made in the particle board sector (including OSB). Investment in the fibreboard sector increased from 1.8 billion roubles in 2014 to 3.2 billion roubles in 2015.

Revenues increased by 10% (to 97 billion roubles) in 2015 in the plywood sector and by 21% (to 50 billion roubles) in the particle board sector (including OSB), but they decreased by 2% (to 21 billion roubles) in the fibreboard sector. The wood-based panels industry accounted for 13% (168 billion roubles) of the total revenue generated by the forest sector in the Russian Federation in 2015 (WhatWood, 2016).

7.3.1 Consumption

Apparent consumption of wood-based panels in the CIS decreased by 4.6% in 2015, to 16.8 million m^3 (table 7.3.1). The consumption of plywood decreased by 13.9%, to 2.0 million m^3, and the consumption of particle board fell by 6.9%, to 8.8 million m^3. In contrast, OSB consumption increased by 35.1%, to 1.67 million m^3, and fibreboard consumption rose by 2.2%, to 4.4 million m^3.

TABLE 7.3.1

Wood-based panel balance, CIS, 2014-2016

(thousand m^3)

	2014	*2015*	*2016f*	*Change (%) 2014-2015*
Production	17,410	18,483	19,537	6.2
Imports	5,954	5,023	5,389	-15.6
Exports	5,783	6,736	7,302	16.5
Apparent consumption	17,581	16,770	17,624	-4.6

Note: f = 2015 Committee on Forests and Forest Industry forecast.
Source: UNECE/FAO, 2016.

7.3.2 Production and capacity utilization

The production of wood-based panels in the CIS increased by 6.2% in 2015, to 18.5 million m^3. The devaluation of the rouble and the high rate of inflation in the domestic market were crucial factors for Russian companies. There was a 3% increase in the production of wood-based panels in the Russian Federation in 2015, to 13.6 million m^3.

Plywood. The production of plywood increased in the CIS by 1.7% in 2015, to 4.0 million m^3. The Russian Federation produced 3.6 million m^3 of this, up by 1.9% (graph 7.3.1; table 7.3.2). The profit margin for Russian plywood mills has increased in recent years, from 3.5% in 2009 to 21.6% in 2015. The most profitable regions for plywood production in the Russian Federation in 2015 were the Novgorod (up by 41%), the Kirov region (up by 28%), the Kostroma region (up by 30%), and the Komi Republic (up by 30%). Investments in basic capital for plywood mills has increased by 170% in the last seven years, amounting to 35 billion roubles in the period 2008 to 2015 (WhatWood, 2016).

Particle board. Particle board production decreased by 2.6% in the CIS in 2015, to 8.9 million m^3. Production by Russian particle board manufacturers (about 40 producers) declined by 7.4%, to 5.7 million m^3 (graph 7.3.1; table 7.3.2). The particle board project in Tomlesdrev, valued at 6 billion roubles, has entered the final stage of construction, including a second particle board mill with a production capacity of 350,000 m³. The project is included in the list of federal priority investment projects and is at the final commissioning stage; the production launch is scheduled for 2016.

GRAPH 7.3.1

Plywood, particle board, fibreboard and OSB production in the Russian Federation, 2011-2015

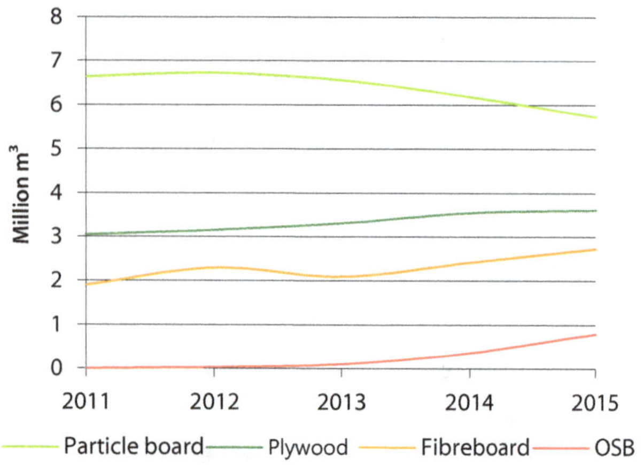

Source: UNECE/FAO, 2016.

OSB. The CIS market for OSB, especially in the Russian Federation, has been one of the most dynamic worldwide in the last decade and has attracted the attention of many investors. Demand for OSB in the subregion doubled every year from 1997 to 2014 (albeit from a low base in 1997). The Russian Federation did not have any OSB production capacity before 2012, and domestic demand was supplied by imports, which peaked in 2013 at 702,000 m³ (valued at $219 million). The situation began to change rapidly in the second half of 2014 when the DOK Kalevala OSB plant neared production capacity, competing in the mid-priced segment of the market with European manufacturers. The slowing of the Russian economy and the devaluation of the rouble helped, leading to a sharp decline in OSB imports. Russian OSB imports fell by 22% in 2014, to 546,700 m³, and total consumption in the domestic market increased by 16%, to 892,000 m³. By 2015, the Russian Federation had three large-scale OSB mills: the DOK Kalevala mill in Petrozavodsk, Karelia, with a production capacity of 300,000 m³; the Kronospan mill in Egorievsk, Moscow region, with a production capacity of 425,000 m³; and the Novovyatsky mill in Kirov, Kirov region, with a production capacity of 100,000 m³. The Hillman OSB mill in Kosterevo, Vladimir region (with a production capacity of 30,000 m³), restarted in the second half of 2015 after an unscheduled shutdown. The Kurgansky Zavod OSB mill in Sychevo, Kurgan region (with a production capacity of 60,000 m³), began test production in September 2015. Kronospan Mogilev, in Belarus (with a production capacity of 600,000 m³), began running in 2014-2015, producing an estimated 450,000 m³ in 2015.

Apparent OSB consumption in the CIS subregion (including imports of 683,000 m³) was 1.4 million m³, up by 19.9% from 2014.

Fibreboard. The production of fibreboard increased sharply (by 16.9%) in the CIS in 2015, to 3.6 million m³. The Russian Federation contributed more than 75% (2.7 million m³) of this total (table 7.3.2).

TABLE 7.3.2

Wood-based panel production, Russian Federation, 2012-2015

(thousand m³)

	2012	*2013*	*2014*	2015	*Change (%) 2014-2015*
Plywood	3,150	3,303	3,540	3,607	1.9
Particle board	6,723	6,555	6,183	5,727	-7.4
Fibreboard	2,291	2,092	2,413	2,722	12.8
OSB	30	101	360	791	119.8

Source: UNECE/FAO, 2016.

Kastamonu Entegre launched its second MDF line in the Russian Federation in early 2016, in Tatarstan. The project started in April 2015 and transport, assembly and commissioning were completed within a year. This €130 million investment increased fibreboard production by 485,000 m³ per year; the plant's MDF production capacity reached 1.05 million m³ and production capacity for laminate flooring under the Floorpan brand reached 40 million m².

Belarus introduced a temporary licensing requirement in 2015 on imports of some types of particle board and fibreboard from outside the common customs territory of the Eurasian Economic Union (Council of Ministries of the Republic of Belarus, 2016). Several new fibreboard mills (including laminate flooring) opened in Belarus In 2014-2015 – in Borisovdrev, Mostovdrev, Gomeldrev and Vitebskdrev. The temporary licensing requirement is intended to reduce imports in favour of domestic products.

7.3.3 Prices

Plywood. Russian producer prices for plywood (averaged for all regions) increased by 23.8% in 2015, to 23,826 roubles per m³, with export prices up by 29%, to 27,255 roubles per m³, and domestic prices up by 12%, to 20,275 roubles per m³ (graph 7.3.2). The highest average prices (domestic and export sales) were in the northwest, where they reached 27,007 roubles per m³ (up by 25% over 2014), while the lowest prices were in Siberia, at 15,884 roubles per m³ (up by 27% over 2014).

GRAPH 7.3.2

Monthly prices for wood-based panels, Russian Federation, 2010-2016

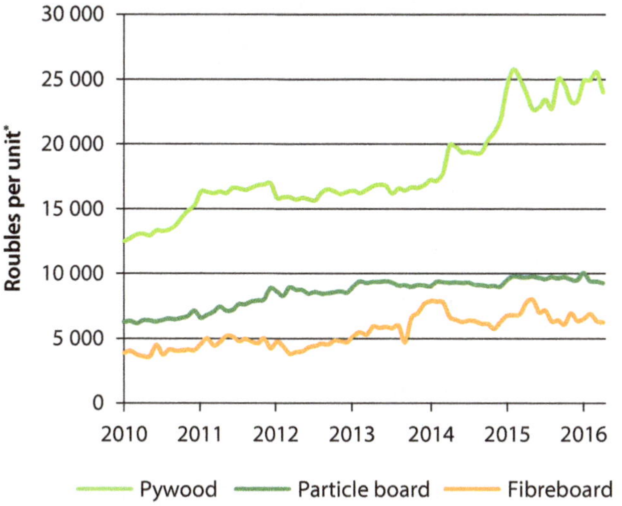

Notes: Sales for exports and domestic markets. *Russian wood-based panel prices (roubles per m³, except fibreboard, which is roubles per 100 m²).
Source: Rosstat, 2016

Particle board. The average price of particle board in the Russian Federation was up by 5.2% in 2015, to approximately 9,607 roubles per m³. The average producer price in the domestic market was 9,743 roubles per m³ (up by 6.5% over 2014), while the average price for exported particle board (primarily going to CIS countries and exchanged in roubles) decreased by 15%, to 7,397 roubles per m³.

The most dynamic growth in prices was in the Russian northwest, where the domestic particle board price was up by 16%, to 12,190 roubles per m³, and the export price was up by 17%, to 10,834 roubles per m³.

Fibreboard. The average price of fibreboard in the Russian Federation rose by 4% in 2015, to 67 roubles per m². The price decreased from 49 to 46 roubles per m² in the Central Federal District, but it increased from 86 to 90 roubles per m² in the Siberian Federal District.

7.3.4 Trade

7.3.4.1 Imports

Plywood. The volume of plywood imported into the CIS subregion decreased by 23.2% in 2015, to 491,000 m³; imports into the Russian Federation were also down sharply (graph 7.3.3).

GRAPH 7.3.3

Fibreboard, OSB, particle board and plywood imports, Russian Federation, 2011-2015

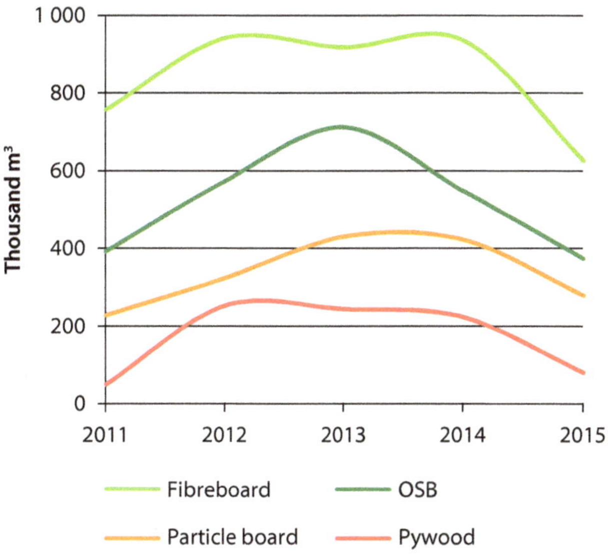

Source: UNECE/FAO, 2016.

OSB. CIS imports of OSB fell significantly (by 23.3%) in 2015, to 658,000 m³, due mainly to currency devaluations in the second half of 2014 and the substitution of imported OSB by domestic production. Canada, Latvia and Romania were the main non-CIS suppliers of OSB to the Russian Federation in 2014, representing about 80% of the total import volume.

Imports of OSB to the Russian Federation fell even more dramatically in 2015, with deliveries from Europe and North America decreasing five-fold, to 107,000 m³. The main reason for the decline was the continued weakness of the rouble, which increased the competitiveness of domestic OSB; moreover, the Belarusian mill, Kronospan-Mogilev, was able to replace some imports from outside the subregion. Canadian OSB disappeared almost completely from the market in the second half of 2015.

Particle board. CIS imports of particle board decreased by 8% in 2015, to 2.08 million m³; particle board imports into the Russian Federation fell by 33.7%, to 280,000 m³. About 62% of all these imports into the Russian Federation were from Poland and Germany.

Fibreboard. CIS fibreboard imports decreased by 16.8% in in 2015, to 1.8 million m³; imports into the Russian Federation fell by 32.8%, to 626,000 m³. China, Germany and Poland were the largest suppliers of fibreboard to the Russian Federation in 2015, with a combined share of about 82%.

7.3.4.2 Exports

Plywood. CIS plywood exports were up by 10.2% in 2015, to just over 2.5 million m³ (graph 7.3.4). Export sales by the Russian plywood industry, which is traditionally export-oriented, reached 2.2 million m³ (up by 12% over 2014), at a value of $991 million.

GRAPH 7.3.4

Fibreboard, OSB, particle board and plywood exports, Russian Federation, 2011-2015

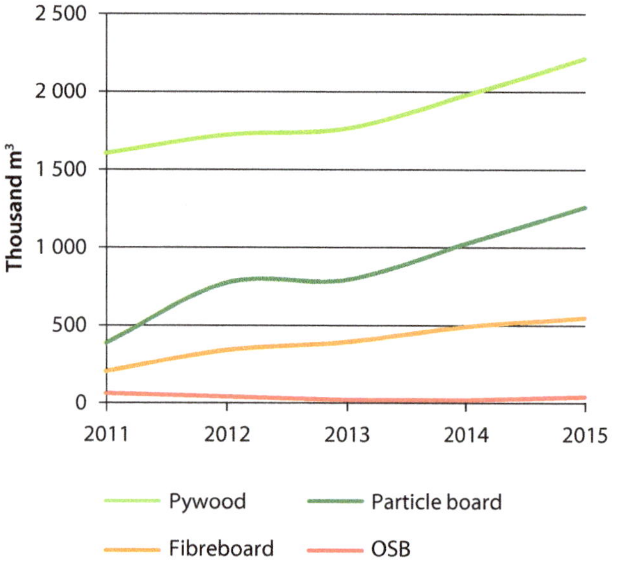

Source: UNECE/FAO, 2016.

Egypt was the largest importer of Russian plywood in 2015, increasing its import volume by 7% to 277,000 m³. The US also increased its Russian plywood imports – by 10%, to 272,000 m³ – to become the second-largest export market for Russian plywood (WhatWood, 2016).

Particle board. CIS exports of particle board increased by 12% in 2015, to just under 2.15 million m³. Russian exports grew by 23.3%, to 1.25 million m³, of which CIS countries imported about 90% (Uzbekistan, for example, accounted for 54.4% of Russian particle board exports).

Fibreboard. Fibreboard exports from the CIS increased by 7.3% in 2015, to 1,052 million m³, of which the Russian Federation accounted for 542,000 m³.

OSB. The CIS exported about 354,000 m³ of OSB in 2015; OSB producers in the CIS have started building sales networks in Central Asia.

7.4 NORTH AMERICA

7.4.1 Consumption

GDP growth slowed significantly in Canada in 2015, to 1.2% (down from 2.5% in 2014), but housing starts still increased by 3.3%, from 189,329 units in 2014 to 195,535 units in 2015. In the US, where GDP growth in 2015 was steady at 2.4%, housing starts increased by 11%, to about 1.11 million units. Apparent consumption of wood-based panels in North America increased by a robust 4.2% in 2015, met largely by strong growth in imports (up by 13.2%, by volume), coupled with slow growth in exports (+2.0%); total wood-based panel production in North America increased by just 1.2% (table 7.4.1).

TABLE 7.4.1

Wood-based panel balance, North America, 2014-2016

(thousand m³)

	2014	*2015*	*2016f*	*Change (%) 2014-2015*
Production	46,039	46,586	46,987	1.2
Imports	13,146	14,877	14,773	13.2
Exports	9,296	9,484	9,763	2.0
Apparent consumption	49,889	51,979	51,997	4.2

Note: f = 2015 Committee on Forests and Forest Industry forecast.
Source: UNECE/FAO, 2016.

The consumption of structural wood-based panels continued its slow recovery in North America in 2015, increasing overall by 4.2% (graph 7.4.1), although an increase in demand for OSB of 6.2% masked a drop in demand for plywood of 1.1%. Consumption increased in all four of the major end-use markets for structural panels – by 5% in the residential construction market, by 2% in the remodelling market, by 2% in the industrial market and by 3% in the non-residential market (graph 7.4.2) (APA, 2016).

GRAPH 7.4.1

Structural panel consumption and housing starts, North America, 2011-2015

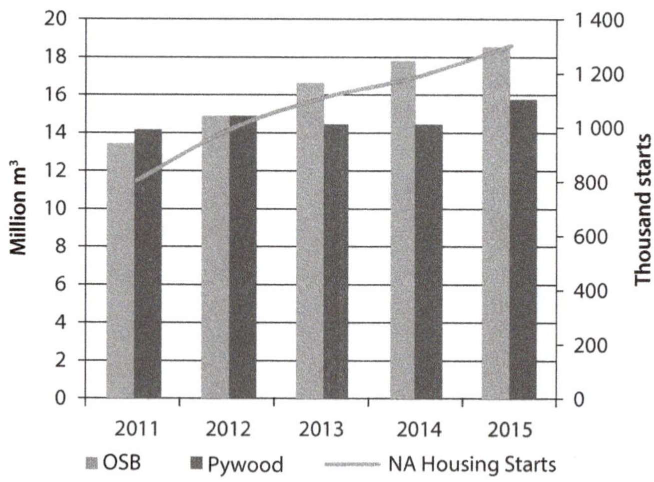

Sources: UNECE/FAO, 2016; APA, 2016.

GRAPH 7.4.2

Four main end-use markets for structural panels, North America, 2015

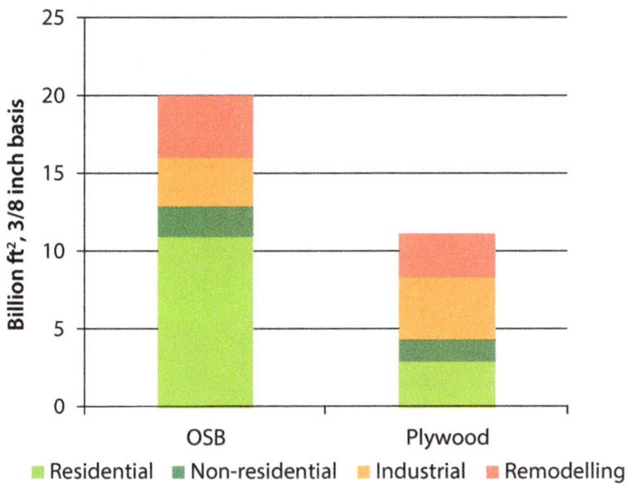

Note: Residential, non-residential and industrial are new construction.
Source: APA, 2016a.

The largest market for OSB in 2015 was residential construction, which consumed 55.2% of total production. The strongest growth in demand for OSB was also in the residential housing sector, with an increase of 8.9% in 2015. Demand for OSB increased by 3% in the remodelling market, by 5.5% in the industrial market and by 5% in the non-residential market. The largest market for plywood use in 2015 was in the industrial sector, which consumed 35.9% of total plywood production. Growth in demand for plywood in 2015 was weak in the industrial (up by 0.5%) and non-residential (up by 0.7%) sectors and negative in the residential housing (down by 5.3%) and repair and remodelling (down by 0.3%) sectors. North American demand for structural panels is expected to increase more strongly in 2016 (by 4.9% overall), with demand projected to grow by 6.1% for OSB and by 2.6% for plywood. All the increase in demand is projected to occur in the US (by 5.6%), with demand in Canada projected to decline slightly (by 0.7%) (APA, 2016). The total consumption of wood-based panels in North America is projected to be flat (up by only 0.03%) in 2015 (UNECE/FAO, 2016).

North American consumption of non-structural panels increased by 3.1% in 2015. Particle board consumption was up by 1.5% and fibreboard consumption grew by 5.3%. With North American housing starts projected to increase modestly in 2016, the production of non-structural wood-based panels is expected to show moderate growth (Composite Panel Association, 2016a, 2016b).

7.4.2 Production and capacity utilization

Production capacity in the North American structural panel subsector increased by 2.2% in 2015, to 38.2 million m^3. This can be attributed to incremental production increases in existing plants because no plant closures or openings were reported in North America in 2015. Capacity utilization in the North American structural panel industry declined from 73% in 2014 to 72% in 2015 but is expected to increase to 74% in 2016. Capacity utilization in the plywood subsector declined from 76% in 2014 to 74% (71% in the US and 90% in Canada) in 2015. Capacity utilization in the OSB subsector was constant overall in North America in 2014 and 2015, at 71% (graph 7.4.3); it was 74% in the US and 66% in Canada in 2015.

GRAPH 7.4.3

Plywood and OSB capacity utilization rates, North America, 2010-2016

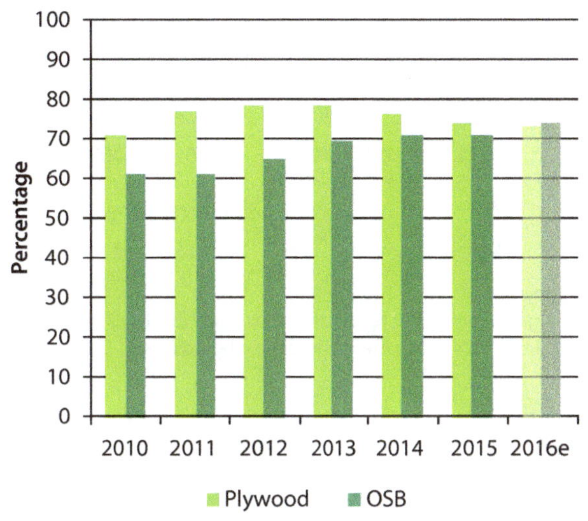

Note: e = estimate.
Source: APA, 2016.

North American production capacity for non-structural panels was flat in 2015, at just over 8 billion square feet (3/4 inch basis) (14.2 million m^3) (Composite Panel Association, 2016b); there was little change in any of the major product categories in 2015. The capacity utilization rate increased in the particle board subsector from 70.7% in 2014 to 72.1% in 2015 and in the MDF sector from 78.8% in 2014 to 81.8% in 2015. Overall, 2015 capacity utilization rates were well below pre-housing crisis levels (Composite Panel Association, 2016a, 2016b).

7.4.3 Prices

Although demand for structural wood-based panels increased in North America in 2015, the capacity utilization rate was unchanged for OSB and declined for plywood. Some interesting trends in prices for these structural panel products were observed as a result. OSB prices bottomed out in the first quarter of 2015 and staged a strong recovery in the final nine months of the year, increasing by 30%; overall, OSB prices increased by 16% in 2015. In contrast, plywood prices, which were relatively stable in the first half of 2015, suffered a significant (14.3%) decline in the second half of the year (graph 7.4.4). Of the non-structural panels, particle board prices were essentially unchanged in 2015 but MDF prices declined slowly, falling by 3.7% over the course of the year (Random Lengths, 2016).

GRAPH 7.4.4

Wood-based panel prices, North America, 2011-2016

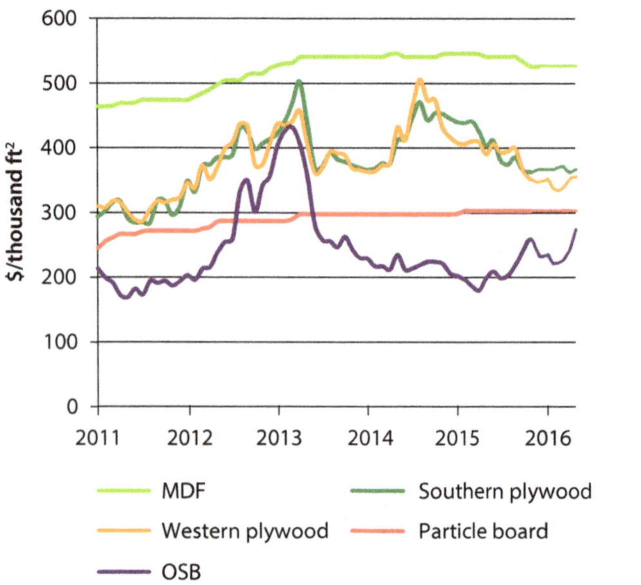

Notes: Western plywood (Coast), ½ inch, CD exterior, (3-ply); OSB (North Central), 7/16 inch; MDF (East) 5/8 inch; particle board (West) 5/8 inch industrial.
Source: Random Lengths, 2016.

7.4.4 Trade

7.4.4.1 Imports

The value of North American imports of wood-based panels increased by 6.5% in 2015, to $5.9 billion (table 7.4.2). Imports to the US grew strongly (by 9.4%), but Canadian imports fell by 8%. Plywood had the largest share of imports to North America (53% of the total value of wood-based panel imports), followed by fibreboard (26%), OSB (17%) and particle board (6%). North American imports increased in 2015 for plywood, fibreboard and OSB and were flat for particle board.

Two import flows of wood-based panels to North America are of particular interest because of their overall size and structure. One is China's domination of plywood imports to the US, with a 52% market share in 2015, followed by Canada (10%), Indonesia (10%) and the Russian Federation (6%). The other is Canada's role as the almost exclusive source of OSB imports to the US; that country accounted for 99.9% of the $917 million worth of US OSB imports in 2015.

TABLE 7.4.2

Value of wood-based panel imports, North America, 2012-2015
($ million)

	2012	*2013*	*2014*	*2015*	*Change (%) 2014-2015*
US					
Plywood	1,906	2,072	2,314	2,681	15.9
Fibreboard	833	971	1,081	1,137	5.2
OSB	772	1,102	936	948	1.3
Particle board	218	251	289	289	0.0
US total	3,729	4,396	4,620	5,055	9.4
Canada					
Plywood	373	370	354	334	-5.6
Fibreboard	447	439	454	410	-9.7
OSB	34	39	36	28	22.2
Particle board	53	57	67	66	-1.5
Canada total	906	905	911	838	-8.0
North America					
Plywood	2,279	2,442	2,668	3,015	13.0
Fibreboard	1,280	1,411	1,535	1,547	0.8
OSB	805	1,141	972	976	0.4
Particle board	271	308	356	355	-0.3
Total	4,635	5,301	5,531	5,893	6.5

Source: UNECE/FAO, 2016.

7.4.4.2 Exports

The value of exports of wood-based panels from North America decreased in 2015 for the second year in a row, dropping by 2.4%, to $2.6 billion; Canada accounted for 71% of the total (table 7.4.3). Structural panels constituted about two-thirds (66%) of 2015 panel export value. The value of wood-based panel exports from North America declined in every category: plywood by 2.1%; fibreboard by 0.2%; OSB by 3.9%; and particle board by 1.9%. The largest markets for US plywood exports in 2015 were Canada (38% by value), Mexico (20%) and Australia (13%). Canada (66% by value), Qatar (15%) and Mexico (12%) were the main markets for US fibreboard, and the main markets for US particle board were Canada (49% by value) and Mexico (33%). Canadian wood-based panel exports went almost exclusively to the US – 91% (by value) of plywood, 96% of fibreboard, 95% of particle board and 94% of OSB exports.

TABLE 7.4.3

Value of wood-based panel exports, North America, 2012-2015

($ million)

	2012	2013	2014	2015	Change (%) 2014-2015
US					
Plywood	431	410	385	346	-10.1
Fibreboard	280	264	246	219	-11.0
OSB	76	90	79	64	-19.0
Particle board	94	99	116	110	-5.2
US total	881	863	826	739	-10.5
Canada					
Plywood	164	214	243	269	10.7
Fibreboard	232	234	263	290	10.3
OSB	884	1,237	1,039	1,009	-2.9
Particle board	201	230	258	256	-0.8
Canada total	1,481	1,915	1,803	1,825	1.2
North America					
Plywood	595	624	628	615	-2.1
Fibreboard	512	498	510	509	-0.2
OSB	960	1,326	1,118	1,074	-3.9
Particle board	295	330	373	366	-1.9
Total	2,362	2,778	2,629	2,565	-2.4

Source: UNECE/FAO, 2016.

7.5 EXTRAREGIONAL INFLUENCES AFFECTING THE UNECE REGION

Japan is the dominant market for plywood outside the UNECE region (Table 7.5.1). Tropical plywood demand in Japan declined in 2014 and 2015, however, despite a surge in late 2013 and early 2014 as consumers rushed to purchase houses before a rise in the consumption tax, which was expected to push up the cost of housing. The Japanese plywood industry was depressed in 2015, despite earthquake reconstruction work and the construction of Olympic venues, with imported tropical plywood also losing significant market share to domestic plywood. Supplies from China, Indonesia and Malaysia declined sharply in 2015 as a weakening yen, log shortages and increased manufacturing costs in Indonesia and Malaysia (the major suppliers) put upward pressure on tropical plywood import prices. Indonesia supplies mainly floor-base plywood, which has a shorter distribution route to final end-users in Japan compared with concrete formwork panels, the major plywood product imported from Malaysia. Consequently, Indonesian suppliers have been able to respond sooner to market signals on housing starts and exchange rates, keeping Japan's inventories at acceptable levels.

Early in 2016, Japan's floor manufacturers were threatening to shift their supply sources to domestic softwoods, but the yen strengthened abruptly, narrowing the price differential between domestic softwood plywood and Indonesian tropical plywood. Demand for Indonesian hardwood plywood recovered quickly in the first quarter of 2016, increasing by 3.4% (by volume) compared with the same period in 2015; demand for Chinese and Malaysian imports remained subdued due to high inventories, however, declining by 22.6% and 12.4%, year-on-year, respectively (ITTO, 2016a). The largest Malaysian plywood mill reportedly began cutting shipments to Japan in April 2016 to restore inventories to acceptable levels (ITTO, 2016a). The Japan Forestry Agency projected that imported plywood demand would increase in the second and third quarters of 2016 but would continue to lose market share to domestic plywood, which is expected to continue to expand in both housing and general construction (ITTO, 2016a).

The Republic of Korea's tropical plywood imports were relatively stable in 2015, with the bulk of the supply coming from China, Indonesia, Malaysia and Viet Nam. Malaysian exports have been affected by the imposition of anti-dumping duties imposed by the Republic of Korea since 2011 on one Sabahan and eight Sarawakian plywood manufacturers. The Korean Wood Panel Association argued successfully that those companies were selling plywood below cost and thus undermining plywood producers in the Republic of Korea. Following a three-month review, authorities in the Republic of Korea decided to extend the duties but at reduced rates from the middle of 2014; import levels from Malaysia have not recovered to previous levels.

Malaysia and Indonesia provided the bulk of global tropical plywood exports in 2015. Malaysia's exports have continued to decline, however, dropping by 26% in 2015, to 1.6 million m^3. Tropical plywood production in Malaysia has become increasingly constrained by the limited availability of raw-material inputs (i.e. peeler logs) to plywood mills and by reduced demand and depressed prices in Japan, the major market. About half Malaysia's tropical plywood exports were shipped to Japan in 2015, followed by the Republic of Korea (9%) and Taiwan Province of China (8%). Malaysian plywood prices have been pushed up by chronic log shortages and rising manufacturing costs, which have been putting upward pressure on prices since 2014.

Although data for 2015 were unavailable for Indonesia's plywood export volumes, exports declined marginally in value terms (by 1.3% compared with 2014), to US$2.34 billion. China and Japan were the major destinations in 2015, each accounting for about one-quarter of Indonesian plywood exports by value; the Republic of Korea, Saudi Arabia, Taiwan Province of China and the US accounted for another 30%. A sharp decline in Japan's imports from Indonesia in 2015 (nearly 20%, by value, compared with 2014) was compensated by significant gains in exports to other major markets, including the US (up by 44%), the Republic of Korea (up by 36%) and Saudi Arabia. Greater awareness of Indonesia's national timber legality assurance scheme, which became operational in 2014, has improved the competitiveness of its plywood exports to the EU.

TABLE 7.5.1

Major importers and exporters of plywood outside the UNECE region, by volume, 2013-2015

(thousand m³)

	2013	2014	2015	Change (%) 2014-2015
Major importers				
Japan	2,603	2,127	1,579	-25.8
Republic of Korea	446	604	603	-0.2
Taiwan Province of China	626	552	n/a	n/a
Singapore	128	140	n/a	n/a
Saudi Arabia	143	117	n/a	n/a
Mexico	129	113	116	2.7
Major exporters				
Indonesia	2,836	2,891	n/a	n/a
Malaysia	3,032	2,781	2,270	-18.4
China	616	763	714	-6.4
Viet Nam	163	222	n/a	n/a

Note: n/a = not available.

Source: ITTO, 2016b; UN COMTRADE, 2016; Global Trade Atlas, 2016.

Note: The statistical annex of the *Forest Products Annual Market Review 2015-2016* is available at: www.unece.org/forests/fpamr2016-annex

7.6 REFERENCES

APA (The Engineered Wood Association). 2016. Market outlook and regional production. Economic Report E82.

Composite Panel Association. 2016a. Profile of the composite panel industry.

Composite Panel Association. 2016b. 2016 North American capacity report.

COMTRADE. 2016. United Nations Commodity Trade Statistics Database. Statistics Division. New York. Available at: http://comtrade.un.org/db

Council of Ministries of the Republic of Belarus. 2016. Available at: www.government.by/upload/docs/file18b8264ea4e71045.PDF

European Panel Federation. 2016. Annual report 2015/2016. Available at: www.europanels.org

Global Trade Atlas. 2016. Available at: www.gtis.com/gta

ITTO. 2016a. Tropical Timber Market Report, 20(12). June 2016. International Tropical Timber Organization (ITTO). Available at: www.itto.int/market_information_service

ITTO. 2016b. Statistical database. International Tropical Timber Organization (ITTO). Available at: www.itto.int/annual_review_output

Random Lengths. 2016. Random Lengths yearbook 2015.

Rosstat. 2016. Federal Service of State Statistics. Moscow. Available at: www.gks.ru

UNECE. 2016. Committee on Forests and the Forest Industry market forecasts. Available at: www.unece.org/forests/fpm/timbercommittee

UNECE/FAO. 2016. TIMBER database. Available at: www.unece.org/forests/fpm/onlinedata

WhatWood. 2016. Russian forest industry in 2016. Available at: www.whatwood.ru

8 PAPER, PAPERBOARD AND WOODPULP

Lead author: *Michel Valois*

Contributing authors:

Eduard Akim,
Bernard Lombard

HIGHLIGHTS

- Paper and paperboard production fell in Europe and North America in 2015 and was flat in the CIS. Capacity closures continued, mainly in the graphic-paper segment.
- Woodpulp production rose in Europe and the CIS, but it fell in North America as a result of longer maintenance periods, integrated closures across the graphic-paper segment, and permanent shut-downs of chemical market pulp capacity.
- Graphic-paper capacity fell by 1.7 million tonnes in the UNECE region in 2015 and is expected to decline by another 1.1 million tonnes in 2016.
- Paper prices stabilized in mid-2015 and rose for newsprint and some graphic-paper grades after years of capacity rationalization. Despite stronger exports from Europe and stronger apparent consumption in North America, prices stagnated for packaging paper and paperboard due to capacity additions.
- Apparent consumption of newsprint in North America fell by 10% in 2015, and production was down by 12.8% (at 5.2 million tonnes). North American newsprint capacity was 5.0 million tonnes in early 2016, down by 1.77 million tonnes from 2014.
- Graphic-paper consumption fell in Europe and North America as a result of continued growth in electronic communication, including via the internet. The newsprint segment experienced the largest reductions in percentage terms, and coated-paper consumption declined as end-users reduced their use of paper or moved to lower-cost uncoated papers.
- The trend of conversion from graphic-paper production to paperboard continued in Europe and the US in 2015 and the first part of 2016.
- Chinese pulp imports have increased in the last decade. A series of large hardwood kraft pulp capacity expansions around the globe caused supply to exceed demand and prices to trend lower in late 2015 and early 2016; prices also trended lower in the same period for softwood kraft grades due to a stronger US dollar and graphic-paper machine closures.
- Prices for recovered paper have increased due to heightened demand and tighter scrutiny by Chinese customs agents to prevent contaminants from entering China. Softwood kraft pulp prices recovered in early 2016 following a correction in mid-2015; hardwood kraft pulp prices fell due to overcapacity in global markets.
- South American chemical market pulp expansions – leading to stronger exports – continued in 2015, particularly among hardwood grades. In general, the lower price of hardwood chemical pulp relative to its softwood counterpart, together with strong pulp demand for tissue production, helped absorb an increase in hardwood kraft capacity.

8.1 INTRODUCTION

The turnaround in the global pulp, paper and paperboard industry that began in 2014 fell apart in 2015 as Asian economies experienced declining export demand, which had a negative impact on packaging paper demand, particularly in China; in addition, trade sanctions against Asian paper exporters reduced growth opportunities. Although pulp shipments were up, values were significantly lower. Overcapacity in the pulp, paper and paperboard segments led to closures and consolidation in 2015 and the first half of 2016.

Cost-reduction strategies and strategic alliances and mergers continued to be implemented among pulp, paper and paperboard companies in Europe and North America in an effort to combat low prices. Such efforts were only partially successful in turning around financial performances, and pulp lines and paper machines were forced to close.

The decision by the US Federal Reserve in December 2015 to raise short-term interest rates caused a sharp rise in the US dollar against most global currencies, which had a negative impact on US exports. This prompted a reversal in the Federal Reserve's plan to further raise rates in 2016, causing an immediate devaluation of the US dollar. This, in turn, helped stabilize pulp, paper and paperboard prices in major global economies by the second quarter of 2016.

Despite years of paper-machine closures, capacity rationalization continued in the paper and paperboard industry in the UNECE region in 2015 following structural changes in the demand landscape and important increases in supply from low-cost producing regions. Significant overcapacity existed in 2015 and early 2016 in the publishing-paper-grades segment of the printing-and-writing subsector as consumers continued to shift to electronic communications. This falling trend led to closures and consolidation, especially in the US. Given the inherent maturity stage of its life cycle, the graphic-paper industry may be ripe for consolidation. The appreciation of the US dollar helped improve financial results for non-US exporters; buyers in markets with weaker currencies continue to require lower import prices, however, because their paper prices remain depressed. More companies in the subsector converted from graphic grades to packaging papers and market pulp output in 2015, primarily in Europe and North America, and a select few in the US have turned their attention to specialty or fluff-pulp production.

The conversion of graphic-paper machines to paperboard slowed in 2015 and early 2016 as the space became increasingly competitive; consolidation is largely over, with the major gains achieved. The main focus of the industry now is on taking advantage of higher standards of living in emerging and underdeveloped markets by investing in personal-care products, such as facial and hygienic tissues, towel and disposable infant and adult diapers, and feminine napkins. Ongoing massive expansions in chemical market pulp mills to meet the fibre requirements for such products continued to fuel improved productivity through the closure of high-cost facilities.

Paper and paperboard production declined in Europe and North America in 2015 and was flat in the CIS (graph 8.1.1).

GRAPH 8.1.1

Production of paper and paperboard, UNECE region, 2011-2016

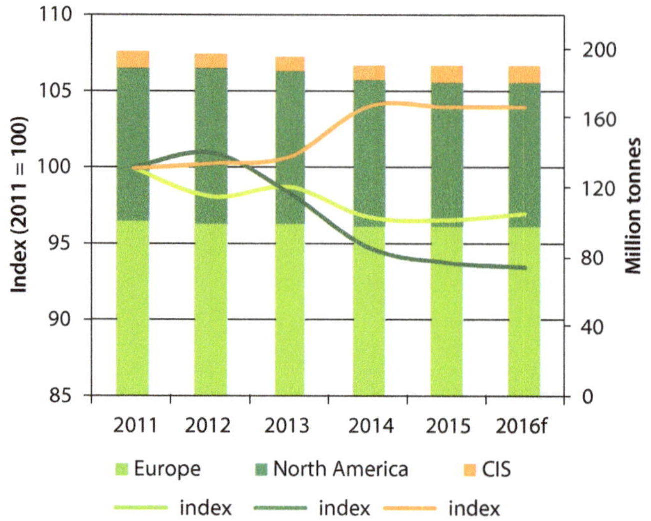

Note: f = 2015 Committee on Forests and the Forest Industry forecast.
Source: UNECE/FAO, 2016.

Corporate strategies continue to focus on cost reductions, establishing new markets, and investing for the future, but all subsectors recognize the need for well-honed logistics for both sourcing raw materials and shipping finished goods to global customers. The complexity of domestic and international trade – such as improved low-cost logistics for selling large quantities to remote markets; high-volume consignments, especially in Europe; just-in-time inventories; documentation; and quality controls for recycled papers entering China – has compelled suppliers to invest in flexible transportation systems. Such investments have increased some costs but also helped ensure timely delivery, improve customer service and reduce inventory carrying costs at both ends. The global pulp, paper and paperboard industry continues to expand into emerging markets, and keeping abreast of the latest production technologies and consumption trends to maximize logistical efficiencies is the key to success. Newer and larger pulp mills continue to displace less-efficient ones, and excess capacity in commodity graphic grades will lead to further closures and industry consolidation. The quest to maintain a "lowest-possible-cost" position will continue to be the focus of commodity grades in each segment of the industry.

In much of Europe, the faltering economic recovery remained a challenge in early 2016. Quantitative easing and a weak euro against the US dollar continued to prop up the economy and favour exports while also causing import costs to rise. In China, GDP growth was 6.9% in 2015 and is expected to remain around that mark in 2016 as exports and domestic consumption remain relatively weak, even with a weaker yuan against the US dollar.

Graphic-paper consumption continued to decline in Europe, Japan and North America in 2015 and into 2016 due to the proliferation of internet-using electronic formats as well as smart-phone and tablet technologies, and the continued trend of end-users using cheaper alternatives to reduce costs. Businesses and governments are pushing for further cost reductions in data manipulation and communication, including traditional mail services, by embracing technology and investing in processes that provide customers with improved, timelier services. With electronic media growing in popularity, the consolidation and closure of printing plants in the newsprint and commercial printing segments continued to make headlines in 2015. Graphic-paper capacity fell by 1.7 million tonnes in the UNECE region in 2015 and is expected to decline by another 1.1 million tonnes in 2016. Four million tonnes of global graphic-paper capacity was indefinitely or permanently removed from production in 2015, and a further drop of 2.7 million tonnes worldwide is expected in 2016 (Valois Vision Marketing, 2016a). Following years of capacity reductions, graphic-paper prices started to recover from dismal and unprofitable levels; however, there is an ever-pressing need to remove inefficient capacity in key markets to further improve the financial performance of the industry globally. North American newsprint capacity was 5.0 million tonnes in early 2016, down by 1.8 million tonnes from 2014. Graph 8.1.2 shows subregional trends in paper and paperboard consumption in 2011-2016.

Outside the UNECE region, pulp capacity continues to increase. In Brazil, a large bleached eucalyptus kraft line with a production capacity of 1.4 million tonnes started up in 2015, followed by 1.5 million tonnes of bleached eucalyptus and softwood kraft in March 2016. In Indonesia, a single mill with two hardwood kraft pulp lines is expected to produce 2.8 million tonnes per year, starting in late 2016. In mature markets such as Europe, Japan and North America, however, market pulp mill closures, integration into tissue and towel operations, and conversions removed 2.1 million tonnes of market pulp capacity in 2015, and another 602,000 tonnes of integrated pulp capacity was permanently or indefinitely removed (Valois Vision Marketing, 2016b).

GRAPH 8.1.2

Apparent consumption of paper and paperboard, UNECE region, 2011-2016

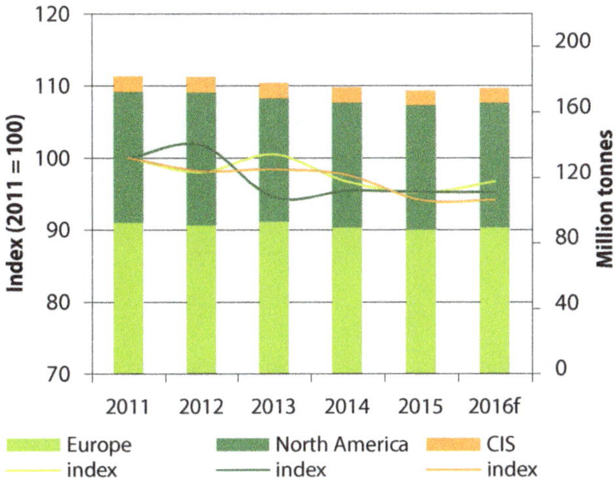

Note: f = 2015 Committee on Forests and Forest Industry forecast.
Source: UNECE/FAO, 2016.

The expansion of woodpulp production in 2011-2016 was concentrated in hardwood grades and in low-cost countries outside the UNECE region. A series of investments in the UNECE region in softwood kraft pulp production, however, saw capacity grow by almost 500,000 tonnes in 2015 and by another 1.3 million tonnes in 2016; this is in stark contrast to 2013-2014, when global softwood kraft capacity stagnated. Significant large-capacity expansion in the bleached hardwood kraft pulp segment – mainly bleached eucalyptus kraft in Brazil – has caused prices to decline, leading to the closure or conversion of relatively high-cost capacity in the UNECE region in the five years to 2016. Specifically, US capacity was permanently closed, while other mills in North America and Europe swung production to softwood kraft grades to take advantage of higher margins. As a result, woodpulp production in the UNECE region trended slightly lower in 2011-2016 (graph 8.1.3).

Aiding the large influx of hardwood kraft in 2015-2016 was the large price differential between it and softwood kraft in global markets, prompting end-users to switch to lower-priced fibre to reduce costs wherever the process and product performance requirements permitted. A slowdown in China's economy in 2015 caused a downturn in pulp prices and the price differential between hardwood and softwood kraft narrowed considerably in the second half of the year. This differential was growing again in mid-2016 as pulp markets recovered; the large incremental hardwood kraft capacity has tended to keep price increases to a minimum relative to those for softwood kraft.

GRAPH 8.1.3

Production of woodpulp, UNECE region, 2011-2016

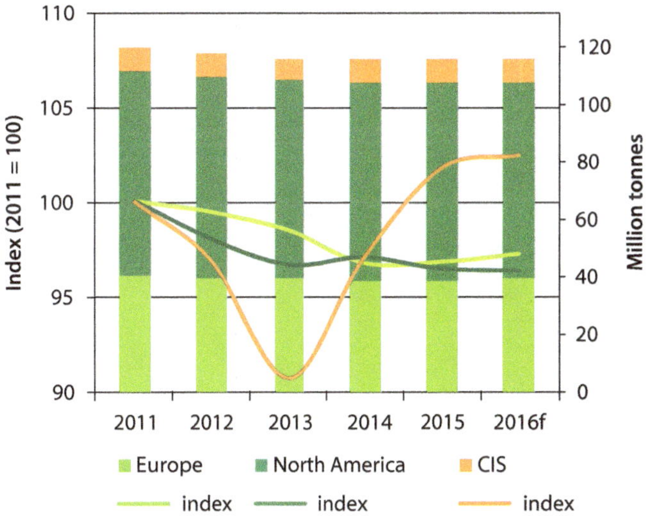

Note: f = 2015 Committee on Forests and the Forest Industry forecast.
Source: UNECE/FAO, 2016.

Capacity rationalization in the pulp and paper subsector continued in Europe, Japan, North America and South America in 2015. Some newsprint machines were closed or converted to packaging grades, and others were converted from paper-grade pulps to dissolving grades. Strong demand in China from the garment industry in particular continued to spur demand for viscose pulps (a subset of dissolving-pulp grades). China continues to impose import duties on dissolving pulps originating from Brazil, Canada and the US; the net result of these duties is that prices have increased in China, even though capacity has expanded in other countries not subject to the duties. Despite the higher prices, some global capacity expansions have been postponed indefinitely (Valois Vision Marketing, 2016c).

Dissolving-pulp demand continued to grow in 2015 and capacity grew in line with this rising demand, allowing prices to increase. In early 2016, however, a major capacity expansion in Brazil targeting the Chinese market added 7% to supply; prices edged lower but still managed to retain two-thirds of the 2015 increases. Fluff-pulp demand also saw solid, sustainable growth in 2015 as standards of living rose in Africa, Asia, the Middle East and South America, aided by higher disposable incomes. Graph 8.1.4 shows overall trends in demand for woodpulp in the UNECE subregions in 2011-2016.

Global prices for softwood kraft pulps started 2015 in decline due to the large price differential between softwood and hardwood kraft pulps and as China's economic growth slowed. Hardwood kraft pulp prices began to erode in mid-to-late 2015 after large capacity additions that exceeded global demand.

Prices for publishing papers were generally flat to weaker in 2015 as supply continued to chase demand downward in all UNECE markets. Prices for coated and uncoated wood-free papers and newsprint began to improve in early 2016 after years of falling demand and overcapacity.

GRAPH 8.1.4

Apparent consumption of woodpulp, UNECE region, 2011-2016

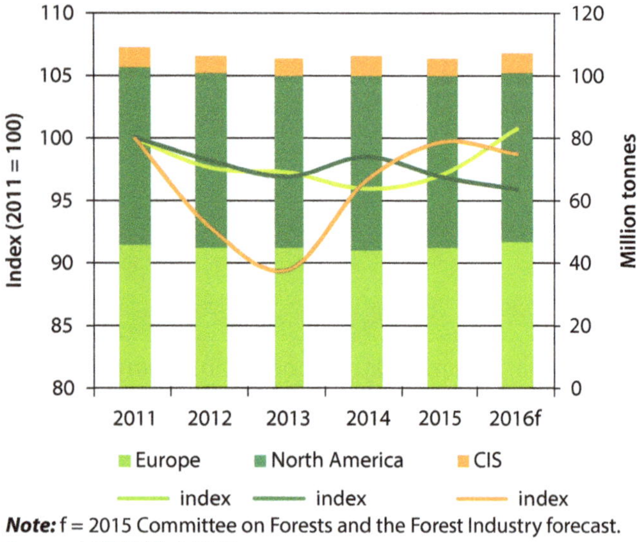

Note: f = 2015 Committee on Forests and the Forest Industry forecast.
Source: UNECE/FAO, 2016.

The global pulp and paper subsector is recovering slowly, aided largely by capacity rationalization. Many difficult reforms have been implemented, including cost-cutting, mergers and divestments, but more are required. Currency fluctuations in 2015 saw global asset valuations decline against the US dollar, causing global trade inequalities and resulting in lower prices. The subsector continues to invest in green technologies (e.g. wood-based biorefineries and biofuels) with the potential to reduce production costs and diversify revenue streams.

Source: Metsäliitto, 2016.

8.2 EUROPE

8.2.1 Paper and paperboard production

The production of paper and paperboard in Europe edged 0.2% lower in 2015 (table 8.2.1). Following years of paper-machine closures, the production of uncoated woodfree papers fell by 1.1% (table 8.2.2) due to closures. The production of uncoated mechanical papers was virtually flat. Excess capacity continued to close and mills were forced to take market–related downtime; production in the newsprint and coated-papers segments declined by 7.8% and 4.5%, respectively, in 2015 (Valois Vision Marketing, 2016a).

TABLE 8.2.1

Paper and paperboard balance, Europe, 2014-2016

(thousand tonnes)

	2014	*2015*	*2016f*	*Change (%) 2014-2015*
Production	98,095	97,887	98,310	-0.2
Imports	54,221	53,219	55,223	-1.8
Exports	62,873	63,109	64,135	0.4
Apparent consumption	89,443	87,998	89,398	-1.6

Note: f = 2015 Committee on Forests and Forest Industry forecast.
Source: UNECE/FAO, 2016.

Weak activity in print advertising due to the ongoing expansion of electronic communication continued to have a negative effect on the publishing subsector. The decline in graphic-paper production is expected to continue and to lead to more mill closures.

The production of sanitary and household papers rose by 2.7% in Europe in 2015, to 7.8 million tonnes, and the production of packaging grades increased by 2.0%, to 49.1 million tonnes, due

TABLE 8.2.2

Production and apparent consumption of paper and paperboard, Europe, 2011, 2014 and 2015

(thousand tonnes)

	Production				Apparent consumption			
	2011	2014	2015	Change (%) 2014-2015	2011	2014	2015	Change (%) 2014-2015
Graphic papers	43,981	38,603	37,228	-3.6	37,665	32,760	31,325	-4.4
Newsprint	9,184	7,485	6,904	-7.8	9,516	7,385	6,855	-7.2
Uncoated mechanical	7,658	6,103	6,106	0.0	6,139	4,769	4,726	-0.9
Uncoated woodfree	9,619	9,648	9,538	-1.1	9,250	8,762	8,568	-2.2
Coated papers	17,520	15,367	14,681	-4.5	12,761	11,844	11,177	-5.6
Sanitary and household papers	7,385	7,632	7,836	2.7	7,562	7,606	7,859	3.3
Packaging materials	45,507	48,136	49,121	2.0	42,941	45,166	44,973	-0.4
Case materials	26,753	28,091	28,698	2.2	27,118	28,048	28,394	1.2
Cartonboard	9,675	10,443	10,710	2.6	7,669	8,384	8,114	-3.2
Wrapping papers	5,062	5,503	5,556	1.0	4,405	4,887	4,586	-6.1
Other papers, mainly packaging	4,016	4,099	4,157	1.4	3,748	3,847	3,878	0.8
Other paper and board	4,419	3,724	3,701	-0.6	4,309	3,911	3,841	-1.8
Total paper and paperboard	101,292	98,095	97,887	-0.2	92,477	89,443	87,998	-1.6

Sources: UNECE/FAO, 2016.

to debottlenecking projects and conversions from newsprint to paperboard. The production of all other grades of paper and paperboard – mainly for industrial and special purposes – fell by 1.4% in Europe due to restructuring, including paper-machine closures.

8.2.2 Paper and paperboard consumption and prices

Apparent consumption of paper and paperboard fell by 1.6% in Europe in 2015 (table 8.2.2). Graphic-paper consumption was down by 4.4%, led by declines of 7.2% in newsprint, 5.6% in coated papers and 2.2% in uncoated wood-free. Apparent consumption of uncoated mechanical papers declined by only 0.9% in 2015 as advertisers swung away from higher-cost coated papers.

The consumption of sanitary and household papers increased by 3.3% in Europe in 2015 and the consumption of packaging materials was down by 0.4% led by growth of 1.2% in case materials and other papers, mainly packaging. The consumption of wrapping papers fell by 6.1%, and the consumption of cartonboard declined by 3.2%.

Prices for most paper and paperboard bottomed out in Europe in 2015, following a downward trend that began in the second half of 2011. After relative stability in 2011, prices for graphic paper declined steadily in 2012-2014. Graphic-paper prices were generally flat in 2015 after years of capacity closures (Valois Vision Marketing, 2016a).

8.2.3 Market and integrated pulp production

Woodpulp production remained stable in Europe in 2015, at 38.4 million tonnes (table 8.2.3). Exports declined by 3.9% and imports decreased slightly (by 0.3%), eve though Europe is used as a location for containerizing pulp arriving from South America (where there is an ongoing shortage of containers) before shipment to Asia. Apparent consumption increased in Europe by 1.2% in 2015.

The production of mechanical pulp fell by 1.4% in 2015 due to the closure of graphic-paper machines. Despite closures, chemical market pulp production increased by 0.1% (Valois Vision Marketing, 2016d).

TABLE 8.2.3

Woodpulp balance, Europe, 2014-2016

(thousand tonnes)

	2014	2015	2016f	Change (%) 2014-2015
Production	38,168	38,194	38,357	0.1
Imports	19,871	19,821	21,445	-0.3
Exports	13,848	13,306	13,416	-3.9
Apparent consumption	44,191	44,709	46,385	1.2

Note: f = 2015 Committee on Forests and Forest Industry forecast.
Source: UNECE/FAO, 2016.

Softwood-based pulp prices declined in Europe through 2015 as a result of a weaker euro against the US dollar, and as supply exceeded demand. Demand was strong in tissue and certain packaging and specialty grades, but it weakened for printing and writing and newsprint due to the increased use of electronic communication means and slower economic activity. Prices began to recover in early 2016 as buyers decided it was time to rebuild depleted inventories and as suppliers redirected tonnage to Asia (primarily China). Despite major expansions in the hardwood kraft segment outside Europe, prices for hardwood kraft pulp, which had fallen in 2014, began to rise in early 2015; prices for hardwood kraft pulps declined, however, towards the end of the year due to excess production and a slowdown in market demand especially from China. As a result, a large price differential – in excess of $200 per tonne at list level – emerged between key softwood and hardwood kraft pulps at the end of 2015 and into 2016. Seeing a clear economic advantage, consumers swung demand from softwood to hardwood pulp (wherever technically possible) to reduce costs.

8.2.4 Use of recovered paper

The use of paper for recycling in Confederation of European Paper Industries (CEPI) [19] countries rose by 0.2% in 2015, to 47.7 million tonnes, and collection increased by 1.4%, to 55.9 million tonnes. Exports of paper for recycling to non-CEPI countries increased by 7.1%, to 10.2 million tonnes; 93% of such exports went to Asian markets.

Paper for recycling comprised 46% of the fibre used for papermaking in CEPI countries in 2015. Woodpulp accounted for another 39.3%, and the remainder (14.7%) comprised non-woodpulp and non-fibrous materials (CEPI, 2016).

8.3 COMMONWEALTH OF INDEPENDENT STATES

8.3.1 Paper and paperboard production and apparent consumption

The production of paper, paperboard and chemical woodpulp rose by 1.9% in the CIS in 2015, to 16.5 million tonnes (table 8.3.1).

TABLE 8.3.1

Production of chemical woodpulp, paper and paperboard, CIS, 2014-2015

(thousand tonnes)

	2014	2015	Change (%) 2014-2015
Chemical wood-pulp	6,518	6,822	4.7
Paper and paperboard	9,721	9,719	0.0
Total	16,240	16,542	1.9

Source: UNECE/FAO, 2016.

[19] Through its 18 member countries (17 EU members plus Norway), CEPI represents 520 pulp, paper and paperboard companies in Europe.

Paper and paperboard production was flat in the CIS in 2015, with no meaningful capacity expansion. The significant devaluation of the rouble against both the euro and the US dollar in recent years facilitated a 1.9% increase in exports of paper and paperboard in the subregion in 2015, but more expensive imports fell by 10.1%. Apparent consumption declined, therefore, by 3.6% (table 8.3.2).

TABLE 8.3.2

Paper and paperboard balance, CIS, 2014-2016

(thousand tonnes)

	2014	2015	2016f	Change (%) 2014-2015
Production	9,721	9,719	9,719	0.0
Imports	2,746	2,470	2,470	-10.1
Exports	3,156	3,215	3,215	1.9
Apparent consumption	9,311	8,975	8,975	-3.6

Note: f = 2015 Committee on Forests and Forest Industry forecast.
Source: UNECE/FAO, 2016.

8.3.2 Chemical woodpulp production and apparent consumption

Chemical woodpulp production in the CIS was 4.7% higher in 2015 than in 2014. The weak rouble reduced subregional imports by 4% and created demand for domestic mechanical pulp, the production of which increased by 5.3% (UNECE/FAO, 2016). Chemical woodpulp exports increased by 8%, resulting in a 2.7% increase in apparent consumption (table 8.3.3).

TABLE 8.3.3

Chemical woodpulp balance, CIS, 2014-2015

(thousand tonnes)

	2014	2015	Change (%) 2014-2015
Production	6,518	6,822	4.7
Imports	251	241	-4.0
Exports	2,063	2,229	8.0
Apparent consumption	4,706	4,834	2.7

Source: UNECE/FAO, 2016.

8.3.3 Russian Federation

8.3.3.1 Production and capacity

The production of pulp, paper and paperboard increased by 3.6% in the Russian Federation in 2015, to 15.9 million tonnes (Rosstat, 2016), in the wake of targeted investment by the private sector, including foreign capital. Contributing to the increase were investments by the Ilim Group in 2013-2015 to expand capacity in northern bleached softwood kraft and white papers, the largest investment in the Russian forest industry in that period; by Arkhangelsk Pulp & Paper in a new semi-chemical pulp line; and by Mondi in additional bleached softwood kraft pulp production at Syktyvkar. The reconstruction and restructuring of the Russian pulp and paper industry continues, aided by export opportunities and the weak rouble, which is stimulating investment in projects that can supply domestic consumption.

8.3.3.2 Imports

The Russian Federation's annual trade deficit in pulp, paper and paperboard fluctuated significantly in the decade 2006-2015. It was $807 million in 2015, down from $2.242 billion in 2014; the reduction was aided by lower imports due to the weak rouble (State Customs Committee, 2016).

8.3.3.3 Exports

Russian exports of pulp, paper and paperboard declined by 14.5% in 2015, to $2.6 billion (State Customs Committee, 2016), despite investments in incremental capacity and a favourable exchange rate as the price of softwood kraft pulp declined in global markets (Valois Vision Marketing, 2015d). Major export destinations in 2015 were China (market pulp and kraft linerboard), India (newsprint), Ireland (market pulp and kraft linerboard) and Turkey (newsprint). Pulp exports exceeded 2 million tonnes in 2015, 66% going to China (Valois Vision Marketing, 2016e).

8.4 NORTH AMERICA

There were several closures of integrated and market pulp operations in North America in 2015, continuing a general trend from the mid-1990s. There were fewer closures in the printing and writing paper, newsprint and paperboard segments than in previous years, but the industry continued to remove high-cost capacity. Because virtually every paper machine resided in an integrated pulp and paper operation, pulp mills were also shuttered. The printing and writing paper and newsprint segments continued to suffer from overcapacity and low prices due to falling paper demand caused by the increasing use of electronic communication means. Newsprint consumption continued to decline due to decreasing newspaper circulations, reduced advertising, lower basis weights, and the impact of the internet. As a result, 425,000 tonnes of newsprint capacity (7.6% of total capacity), as well as 154,000 tonnes of printing and writing capacity (less than 1% of total capacity), were idled indefinitely or closed permanently in North America in 2015 (Valois Vision Marketing, 2015b).

Overcapacity continued in the newsprint subsector in 2015 due to a 10.8% decline in North American demand (to 3.3 million tonnes). Consumption by daily newspapers fell by 12.7% in 2015 (after a decline of 12.4% in 2014), and consumption in non-newspaper applications, such as flyers and inserts, fell by 2.4% (after an increase of 2.5% in 2014) (Valois Vision Marketing, 2016d).

Tissue production in North America edged 0.8% higher in 2015 following a major wave of investment in 2012-2015, virtually entirely in the US, which increased capacity by 8.7%, to 9.28 million tonnes. Tissue capacity is forecast to increase further in North America in the short term, driven by increased demand for private brands, product-line extensions, and premium hygienic tissue. After a review that began in June 2015, the US International Trade Commission renewed a five-year tariff of 112.64% on the import of Chinese tissue products, defined as cut-to-length sheets of tissue paper with a basis weight not exceeding 29 g/m^2 and a width of no less than 1.27 cm. The US Department of Commerce put the tariff in place in 2005 after its investigation determined that 12 producers and exporters sold certain Chinese-made tissue products to the US at less than their fair value, with a margin of 112.64%.

Canadian market pulp exporters benefited from a weaker currency against the US dollar in 2015, which initially increased margins for producers; much of the currency advantage disappeared, however, due to weaker prices across chemical and mechanical grades in global markets. The stronger US dollar also attracted imports of printing and writing paper, causing several US mills to lose market share rapidly in a segment where overall demand was already declining. Affected parties petitioned the US International Trade Commission and the US Department of Commerce, which in turn launched an investigation into imports of cut-size office and copy paper. A series of anti-dumping duties was imposed on imports of such papers from certain producers in several countries, including Brazil, China and Indonesia. In another case (now under review by the US International Trade Commission), anti-dumping duties were imposed on imports of Canadian super-calendared paper in 2015.

As for much of the last 15 years, domestic North American demand for graphic paper declined in 2015 as a result of lower advertising budgets for print advertising and growth in electronic media for data transmission and information dissemination. Further paper-machine closures are anticipated and restarts seem unlikely. This pattern is being replicated in mature markets worldwide.

8.4.1 Production and apparent consumption

North America's apparent consumption of paper and paperboard, which had been trending lower for years, fell only slightly (by 0.2%) in 2015, to 75.9 million tonnes (table 8.4.1). This steadying followed years of capacity rationalization and industry consolidation and was aided by lower imports, stronger tissue consumption, and the rebuilding of depleted inventories by graphic-paper consumers ahead of price increases not seen in many grades since 2010. There was sustained demand and profit growth in the paperboard subsector following industry consolidation and capacity rationalization and as a result of growth in online shopping, which boosted the consumption of packaging and shipping cases.

TABLE 8.4.1

Paper and paperboard balance, North America, 2014-2016

(thousand tonnes)

	2014	2015	2016f	Change (%) 2014-2015
Production	83,868	82,984	82,709	-1.1
Imports	12,664	12,155	12,139	-4.0
Exports	20,479	19,216	18,926	-6.2
Apparent consumption	76,053	75,923	75,921	-0.2

Note: f = 2015 Committee on Forests and Forest Industry forecast.
Source: UNECE/FAO, 2016.

The production of paper and paperboard fell by 1.1% in North America in 2015 as graphic-paper capacity was permanently shut; expansions in the tissue and specialty paper and packaging segments paled in comparison with the 580,000 tonnes of graphic-paper capacity idled in 2015.

North American paper and paperboard imports declined by 4% in 2015 due to the imposition of anti-dumping duties on cut-size office and copy paper imports (as mentioned above). Exports fell by 6.2% due to permanently idled capacity and a stronger US dollar.

The production of graphic paper in North America fell by 5.7% in 2015 (table 8.4.2) as capacity was permanently removed due to falling demand and competition from imports, continuing a decline that has cut almost 20% from production since 2010. The production of packaging materials, on the other hand, grew by 0.7% as US manufacturing improved and online shopping propelled demand for case materials. There were significant capacity closures in the production and converting segments in 2011-2014 as a result of industry mergers and acquisitions; prices were lower in virtually all paperboard subcategories in 2015, however, due to increased production and competition from imports. A series of newsprint conversions and debottlenecking projects added 1.1 million tonnes to paperboard production in 2014-2015, following a similar increase in 2013 (Valois Vision Marketing, 2016d).

The production of newsprint fell by 12.8% in North America in 2015, driven by capacity rationalization, including conversions to packaging grades for which margins tended to be higher. Uncoated mechanical paper production rose by 3.9% as advertisers and other end-users swung away from more-expensive coated papers, the production of which fell by 10.4%. The production of uncoated wood-free fell by 1% in 2015 and the production of sanitary and household papers rose by 1.3%.

North America's apparent consumption of graphic papers dropped by 3.8% in 2015, to 22.2 million tonnes (table 8.4.2 and graph 8.4.1). Apparent consumption of newsprint declined by 10%, to 3.4 million tonnes (it has fallen by 7.7 million tonnes since January 2004). Apparent consumption increased by 3.3% for uncoated mechanical papers, fell by 7.4% for coated papers and by 1.2% for uncoated wood-free, and grew by 1.4% for sanitary and household papers. The apparent consumption of packaging materials in North America increased by 1.6% in 2015, to 44.6 million tonnes.

TABLE 8.4.2

Production and apparent consumption of paper and paperboard, North America, 2011, 2014 and 2015

(thousand tonnes)

	Production				Apparent consumption			
	2011	2014	2015	Change (%) 2014-2015	2011	2014	2015	Change (%) 2014-2015
Graphic papers	28,080	24,343	22,961	-5.7	26,076	23,125	22,236	-3.8
Newsprint	7,346	5,939	5,179	-12.8	4,675	3,822	3,440	-10.0
Uncoated mechanical	4,438	3,433	3,566	3.9	4,506	3,470	3,586	3.3
Uncoated wood-free	9,119	8,538	8,450	-1.0	9,047	8,819	8,716	-1.2
Coated papers	7,177	6,433	5,766	-10.4	7,848	7,013	6,494	-7.4
Sanitary and household papers	7,336	7,427	7,521	1.3	7,442	7,540	7,646	1.4
Packaging materials	49,023	50,352	50,728	0.7	42,207	43,885	44,565	1.6
Case materials	32,888	33,949	34,424	1.4	28,294	29,796	30,404	2.0
Carton	7,285	12,158	12,030	-1.1	5,757	10,539	10,411	-1.2
Wrapping papers	3,140	1,854	1,969	6.2	2,443	1,161	1,450	24.8
Other papers, mainly packaging	5,710	2,391	2,305	-3.6	5,713	2,388	2,300	-3.7
Other paper and board	4,049	1,746	1,774	1.6	3,996	1,504	1,476	-1.8
Total paper and paperboard	88,488	83,868	82,984	-1.1	79,721	76,053	75,923	-0.2

Source: UNECE/FAO, 2016.

GRAPH 8.4.1

Apparent consumption of paper and paperboard, North America, 2011-2015

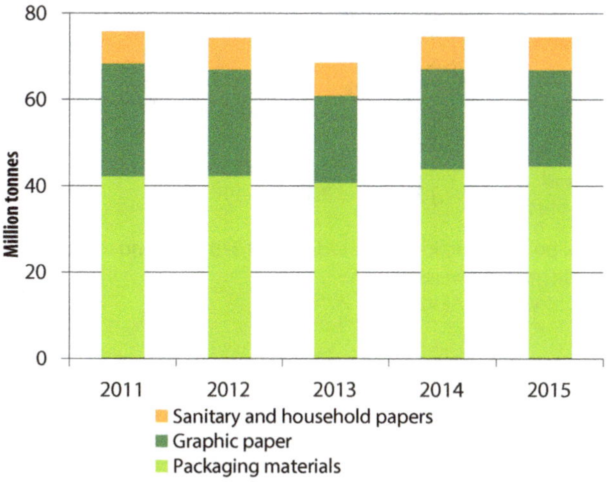

Source: UNECE/FAO, 2016.

The production of chemical woodpulp fell by 1.3% in North America in 2015 (graph 8.4.2) as printing and writing paper production capacity was cut in response to falling demand; reductions in production included permanent closures, longer planned and unplanned maintenance periods, and temporary and indefinite downtime (Valois Vision Marketing, 2016d). Apparent consumption fell by 2.5% as high-cost paper machines were closed permanently and pulp capacity was converted to value-added grades of fluff, dissolving and nano-crystalline cellulose pulps, which generate lower outputs due to their highly refined natures (Valois Vision Marketing, 2015d).

Significant merger and acquisition activity in the North American paperboard subsector in 2014-2015 caused a drop in integrated capacity, which led to higher efficiencies through cost-cutting measures and a tighter supply–demand balance. As a result, prices tended to stabilize or bottom out, allowing for improved financial results. Prices in the subsector continued to be challenged in 2015, however, by competition from imports and increased capacity due to conversions of newsprint-paper machines to paperboard production.

GRAPH 8.4.2

Production of chemical woodpulp, North America, 2011-2015

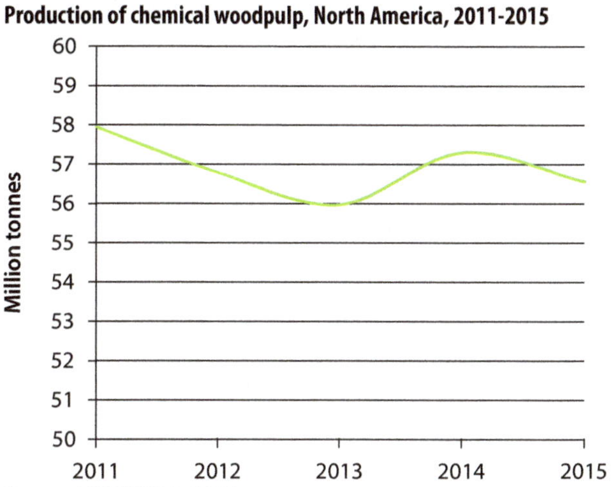

Source: UNECE/FAO, 2016.

The chemical market pulp subsector saw modest growth in export markets in 2015, led by Asia and especially China, where major investments in incremental capacity were again noticeable (Valois Vision Marketing, 2016d). Demand was broad-based in 2015, stemming from investment in tissue and specialty-paper machines as well as building-material applications, such as fibre-cement board.

8.5 EXTRAREGIONAL INFLUENCES AFFECTING THE UNECE REGION

8.5.1 South America

8.5.1.1 Brazil

Pulp capacity expansions in South America added 1.5 million tonnes of bleached eucalyptus market pulp capacity in 2015, an increase in global bleached hardwood kraft capacity of 4.1%. Planned further expansions will add 2.8 million tonnes of bleached chemical market pulp capacity in Brazil by the end of 2018, an increase of 20.2% compared with the country's existing capacity (Valois Vision Marketing, 2016f).

Brazilian pulp production was 17.2 million tonnes in 2015, an increase of 4.6% resulting from the start-up of yet another new pulp line. Paper and paperboard production declined by 0.5%, however, due to ongoing lacklustre economic conditions.

Brazil exported more than 11.5 million tonnes of pulp in 2015, which was 67% of the country's total production (table 8.5.1). The export volume was up by 8.6% compared with 2014, when 10.6 million tonnes – 64.5% of that year's production – were exported (Ibá, 2016).

TABLE 8.5.1

Woodpulp balance, Brazil, 2014-2015

(thousand tonnes)

	2014	*2015*	*Change (%) 2014-2015*
Production	16,461	17,214	4.6
Exports	10,614	11,528	8.6
Imports	416	407	-2.2
Apparent consumption	6,263	6,093	-2.7

Source: Ibá, 2016.

Unlike pulp, most paper and paperboard produced in Brazil is consumed internally. Exports accounted for 19.9% of production in 2015, up from 17.8% from 2014 (table 8.5.2).

TABLE 8.5.2

Paper and paperboard balance, Brazil, 2014-2015

(thousand tonnes)

	2014	2015	Change (%) 2014-2015
Production	10,397	10,343	-0.5
Domestic sales	5,714	5,453	-4.6
Exports	1,846	2,058	11.5
Imports	1,262	866	-31.4
Apparent consumption	9,813	9,151	-6.7

Source: Ibá, 2016.

8.5.1.2 Chile

Chile's exports of pulp, paper and paperboard fell by 7.1% in 2015 (table 8.5.3), due in part to slower economic growth in China, which, along with incremental supply growth in hardwood kraft pulp in excess of demand, contributed to declines in woodpulp prices. Chile's aggregate pulp exports fell by 7.6% and paperboard exports fell by 4% due to lower demand in Latin America.

TABLE 8.5.3

Pulp, paper and paperboard exports, Chile, 2014-2015

(thousand tonnes)

	2014	2015	Change (%) 2014-2015
Bleached radiata kraft	2,138	1,869	-12.6
Bleached eucalyptus kraft	2,091	2,028	-3.0
Unbleached radiata kraft	441	420	-4.8
Newsprint paper	54	54	0.0
Paperboard	499	479	-4.0
Total	5,223	4,850	-7.1

Source: Infor, 2016.

8.5.2 Asia

8.5.2.1 China

China's economic slowdown continues, with fewer exports but also lower commodity prices. The Chinese government, which initiated social and economic reforms in 2014, attempted to further stimulate the economy in 2015 and 2016 with investments in infrastructure, which, in turn, stimulated private investment. The government also abandoned the one-child policy, which should help in avoiding some of the negative economic effects of an ageing population.

China's pulp production rose by 1% in 2015, with less market downtime than in 2014 (table 8.5.4). A surge in bleached hardwood pulp capacity in 2015 caused prices to drop in early-to-mid 2016 (Valois Vision Marketing, 2016e).

TABLE 8.5.4

Production and apparent consumption of pulp, paper and paperboard, China, 2015

(thousand tonnes)

	Production		Apparent consumption	
	2015	Change (%) 2014-2015	2015	Change (%) 2014-2015
Pulp	79,084	1.0	97,310	3.7
Paper and paperboard	107,100	2.3	103,520	2.8

Note: The pulp production figure includes 63,376 thousand tonnes of pulp made from recovered paper.
Source: China Paper Association, 2016.

Apparent consumption of woodpulp rose by 3.7% in China in 2015, driven mainly by growth in tissue and paperboard production. China's overall paper and paperboard production increased by 2.3% following the start-up of several new linerboard, corrugating medium, and specialty-paper machines (Valois Vision Marketing, 2016e). The country's apparent consumption of paper and paperboard rebounded in 2015 by 2.8%, the result of stronger demand in the graphic-paper and cardboard segments. Significant investment continues to be made in the industry, with tissue machines leading the field. Several tissue-machine expansions that were scaled back from previously announced, overambitious plans were implemented in 2015 and will continue until 2018.

China's kraft pulp imports reached a record 15.8 million tonnes in 2015, a 10.5% increase over 2014, while mechanical pulp imports were 1.73 million tonnes, a 14.9% increase (table 8.5.5). Total pulp imports grew by 10.4%, to 19.8 million tonnes.

TABLE 8.5.5

Pulp imports, China, 2013-2015

(thousand tonnes)

	2013	2014	2015	Change (%) 2014-2015
Kraft	13,571	14,277	15,783	10.5
Mechanical	1,388	1,508	1,732	14.9
Dissolving	1,804	2,086	2,247	7.7
Other	87	92	76	-17.4
Total	16,850	17,963	19,838	10.4

Source: China Customs Bureau, 2016.

Despite slower economic growth, China continues to source large volumes of recovered paper to feed its growing papermaking industry; recovered paper represents by far the largest source of fibre in the country's paper and paperboard industry. China's recovered-paper imports fell by 3.7% in 2015, however, to 26.5 million tonnes, as customs officials continued their enforcement of quality controls (table 8.5.6).

TABLE 8.5.6

Recovered paper imports, China, 2014-2015

(million tonnes)

	2014	Share (%) of total	2015	Share (%) of total
Imports by China	27.5	100	26.5	100
Of which from the US	13.6	47.1	14.7	55.5
US exports	19.1	100	19.5	100
Of which to China	13.6	68.1	14.7	75.4

Source: China Customs Bureau, 2016; US Census Bureau, 2016.

Source: UPM, 2016.

An estimated 55% of Chinese recovered-paper imports were sourced from the US in 2015 (US Census Bureau, 2016), up from 47% in 2014. The increase was attributed to an increase in Chinese paper and paperboard production of 2.3%.

8.5.3 Dissolving pulp demand in emerging markets

Global demand for dissolving pulp continues to grow, aided by the development of new end uses and rapid economic growth in emerging markets. The increased consumption of dissolving pulp is being driven by consumers in emerging markets looking to improve their standards of living. Applications include personal-care products, packaging, detergents, foods, textiles and car parts. There has been very strong growth in rayon, a major subcategory of dissolving pulp, as clothing manufacturers abandon expensive natural fibres such as cotton.

Dissolving-pulp capacity expansions in Brazil, Canada, China, Sweden and the US have created a surplus, putting pressure on prices in global markets. Strong demand in China and a swing from dissolving pulp production to bleached hardwood kraft and fluff pulp allowed a modest pricing recovery in 2015. Dissolving pulp capacity in December 2015 was estimated at 7.7 million tonnes, while demand was 6.7 million tonnes.

China's imports of dissolving pulp increased by 7.7% in 2015, to 2.2 million tonnes (table 8.5.5).

Source: UNECE/FAO, 2014.

China imposed import duties on dissolving-pulp imports from Brazil, Canada and the US in 2014; these duties are in place for a minimum of five years and could be extended following a review of market conditions and domestic producers. Despite the duties, global capacity expanded further and prices fell to around $800 per tonne, down from $860 when the duties were imposed. By mid-2015, prices had recovered to $860-$870, but growth in excess capacity in early 2016 caused another price drop, to $840.

As the dissolving-pulp market expands and a select few pulp producers delve into nano-cellulosic fibres, the quest to develop niche markets will be the modus operandi for improving financial results. Nano-cellulose pulp is a highly refined material that can be used to strengthen products such as paper, plastics and other composites and also as an improved barrier film for food packaging. This industry is in its infancy, and several companies worldwide are involved in research into, and the development of, new applications.

Note: The statistical annex of the *Forest Products Annual Market Review 2015-2016* is available at: www.unece.org/forests/fpamr2016-annex

8.6 REFERENCES

Ibá, 2016. Available at: www.bracelpa.org.br/bra2/?q=en

CEPI. 2016. CEPI preliminary statistics 2015. Confederation of European Paper Industries (CEPI). Available at: www.cepi.org

Infor. 2016. Chilean customs data. Available at: www.infor.cl

China Paper Association. 2016. Available at: www.enchinappi.org

China Customs Bureau. 2016. Pulp imports. Available at: www.chinacustomsstat.com

Rosstat. 2016. Available at: www.gks.ru/wps/wcm/

connect/rosstat_main/rosstat/en/main/

State Customs Committee, 2016. Pulp, Paper, Board magazine, PPB-express, PPB Exports, PPB Imports, RAO Bumprom

UNECE/FAO. 2016. Timber database 2014-2015. Available at: www.unece.org/forests/welcome.html

US Census Bureau. 2016. Available at: www.census.gov

Valois Vision Marketing. 2016a. Paper machine closures database. Available at: www.valoisvision.com

Valois Vision Marketing. 2016b. Pulp mill closures & conversion study 2016. Available at: www.valoisvision.com

Valois Vision Marketing. 2016c. Dissolving capacity & expansions. Available at: www.valoisvision.com

Valois Vision Marketing. 2016d. Flash report. Available at: www.valoisvision.com

Valois Vision Marketing. 2016e. China study. Available at: www.valoisvision.com

Valois Vision Marketing. 2016f. Supply-demand analysis. Available at: www.valoisvision.com

K10777
62t

9 WOOD ENERGY

Lead author: ***Francisco X. Aguilar***

Contributing authors:

Karen Abt,
Branko Glavonjić,
Eugene Lopatin
Warren Mabee

HIGHLIGHTS

- Recent data for the EU28 show that primary energy production from "solid biofuels (excluding charcoal)" decreased by 3% in 2014 compared with 2013, to 3,591 PJ. Solid biofuels accounted for 43.8% of primary energy production from renewable sources.
- Demand for wood energy feedstock is increasing in the CIS as wood energy consumption (including pellets, briquettes and chips) continues to grow in the CIS and neighbouring countries. Wood pellet production in the Russian Federation was 973,000 tonnes in 2015, the highest since 2009.
- Wood energy consumption in North America was estimated at 2,725 PJ in 2015. Consumption increased in Canada by about 8% in 2015, due partly to the harsh winter there. On the other hand, wood energy consumption declined by 7.6% in the US, led by a 22% contraction in residential wood energy consumption, likely a result of lower fossil fuel prices.
- In Europe and North America, industrial and residential (premium) pellet prices declined year-on-year in 2015, driven partly by oversupply and partly by lower demand linked to mild winter temperatures. There is uncertainty in the market because installed wood pellet manufacturing capacity is growing but the extent of additional demand is unknown. Potential changes to eligibility rules and financial incentives for renewable energy are another source of uncertainty.
- North America exported 6.2 million tonnes of wood pellets in 2015, about 5.8 million tonnes of which went to the EU28; the UK is the EU28's largest importer of industrial wood pellets. The devaluation of local currencies in CIS countries supported an increase in export sales there by wood-processing companies, including of wood energy products.
- The production of all woodfuels except charcoal increased in the western Balkans in 2015, reaching a value of 214.2 PJ. Firewood production in the subregion was 19.9 million m^3, a record high, and wood pellet production was 909,600 tonnes. An analysis of the woodfuel trade flows of western Balkan countries in 2015 showed that about 81% of woodfuel was exported beyond the subregion, mostly to Italy.
- The extent to which the adoption of the Paris Agreement will support an expansion of wood energy markets depends on many external factors, including the frameworks adopted for carbon accounting.
- Small-scale, highly efficient power-and-heat-generation units using woodfuels present opportunities for development, and some seem close to commercial feasibility.

9.1 INTRODUCTION

The availability of information on wood energy[20] continues to improve, particularly for commoditized woodfuels. Wood energy consumption and production vary in the UNECE region because demand is strongly affected by weather and the prices of competing energy sources. There has been an increase in wood energy in the power-and-heat sector in the EU28 and North American subregions in the last ten years, with prospects for a similar trend in the CIS. The implementation of renewable energy targets and financial incentive programmes, and improvements in energy conversion efficiency, has helped drive this trend. Across the UNECE region, other factors influencing wood energy markets include requirements to address bioenergy sustainability and the role of wood energy in reducing greenhouse gas (GHG) emissions. Programmes developed to ensure standardized wood pellet quality and sustainability exemplify proactive multistakeholder efforts to deal with these potential concerns. Government action requiring the certification/verification of wood-fibre sources, the establishment of GHG accounting frameworks, and updated renewable energy targets can significantly influence commoditized wood energy markets.

Source: A. Merkulov, 2016

9.2 EUROPE

9.2.1 Consumption and production

Recent data show that the primary production of "solid biofuels (excluding charcoal)" in the EU28 decreased by 3% in 2014 compared with 2013, to about 3,591 PJ[21] (Eurostat, 2016b). Nevertheless, primary energy production from solid biofuels increased by 30% in the EU28 in the ten years from 2005 to 2014 (graph 9.2.1). Solid biofuels accounted for 43.8% of primary energy production from renewable sources in 2014 – they were the main source of renewable energy in the EU28 in that year, followed by hydro (16.5%) and wind (11.1%). Overall, primary energy production from renewables increased by 1.6% in 2014, the lowest annual increase in 15 years. EU28 imports of solid biofuels have increased three-fold since 2005 (Eurostat, 2016a).

GRAPH 9.2.1

EU28 total primary energy production from solid biofuels, and share of imports, 2005-2014

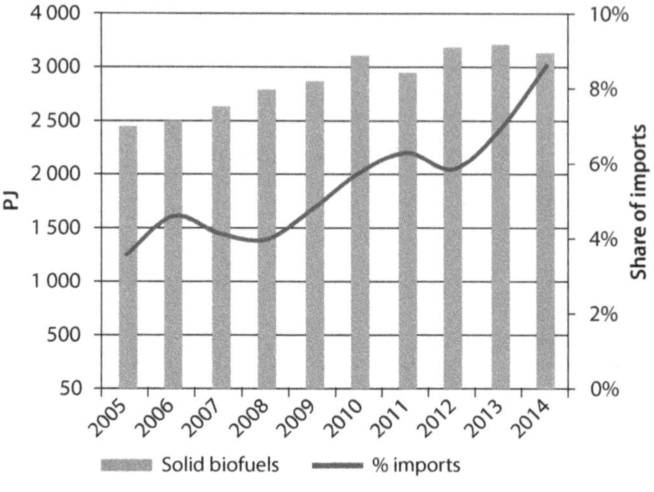

Note: Excluding charcoal.
Source: Eurostat, 2016c.

An estimated 13.1 million tonnes of wood pellets were produced in the EU28 in 2014, an increase of more than 1 million tonnes compared with 2013 (Eurostat, 2016bc). Germany was the largest producer of pellets in the EU28 in 2014, accounting for 16% of production, followed by Sweden (12%), Latvia (10%) and France (9%). The EU28 contributed about half of global wood pellet production in 2014 (Calderon et al., 2016).

The EU28 accounted for 74% of global wood pellet consumption in 2014, inclusive of uses for electricity, heat and combined-heat-and-power (CHP) generation (Calderon et al., 2016). EU28 consumption of premium pellets [22] for residential heating increased to 11.7 million tonnes in 2015, about 7% higher than in 2014 (Calderon et al., 2016). Italy continued to be the largest national market for residential pellets in the EU28, consuming some 3 million tonnes in 2015. Two major factors in Europe that affected demand for wood pellets for heating in 2015 were the mild winter and the continued downward price trend for fossil fuels (Argus Media, 2016b). Sales of wood pellet boilers and stoves declined: sales of boilers were lower in Austria, France and Germany, and stove sales fell by 42,000 units in 2015, year-on-year, to 357,000. Italy is the largest market, accounting for about 90% of European sales, although the French market continues to grow. Prospects are more promising for stoves than for boilers due to their lower capital cost and the opportunity to replace logwood stoves with more efficient and convenient wood pellet stoves (Argus Media, 2016b).

Results for the industrial wood pellet market were mixed in 2015: consumption in Europe increased to an estimated 7.8 million tonnes (Calderon et al., 2016), but the industry also faced lingering uncertainties, with German Pellets, one of the subregion's largest wood pellet manufacturers, filing for

[20] Charcoal is excluded except where otherwise indicated.

[21] One PJ equals about 105,000 m^3 of wood or about 50,000 oven dry tonnes of wood fibre (higher heating value).

[22] Premium pellets are intended for applications where a low ash content (≤ 1%) is desirable, with moisture content ≤8% and fine particles ≤0.5%. For more details see the Pellet Fuels Institute standards at www.pelletheat.org.

insolvency in February 2016, and its US subsidiary, Louisiana Pellets, also filing for bankruptcy protection. Another subsidiary, Hot'ts Holzpellets (with plants in Germany and Austria), has been sold, and agreements are in place to sell remaining plants (Argus Media, 2016a,b). Concerns remain, too, about the sustainability of solid biomass fuels.

Wood energy consumption reached a record high of 161.1 PJ in the western Balkans in 2015. The consumption of wood pellets increased by about 122,000 tonnes in 2015, year-on-year, with growth most pronounced in Bosnia and Herzegovina, Serbia and Slovenia. The increase in subregional consumption enabled many producers to sell pellets that had been stockpiled in 2014 due to increased competition in the Italian market (the main destination for wood pellets from the western Balkans) from Canadian and US producers (Glavonjić, 2016).

The residential and power-and-heat sectors are the two largest consumers of wood energy in the western Balkans. Households depend heavily on firewood, mainly for heating, with the share of total heating ranging from 53% in Croatia to 92% in The former Yugoslav Republic of Macedonia. There is also a reported trend towards the increased use of wood pellets for residential heating. For example, some 10,000 households in two large towns (Sarajevo and Banja Luka) in Bosnia and Herzegovina disconnected from the district heating system network at the beginning of the 2015-2016 heating season, instead relying on residential boilers using wood pellets. There was a rapid increase in wood chip consumption in district heating systems in Bosnia and Herzegovina – about 45,000 tonnes of wood chips were consumed in 2015 by five district heating systems that had substituted light oil with wood chips. Other towns in Bosnia and Herzegovina announced plans in 2016 to substitute light oil with wood chips in district heating systems, which could increase consumption to over 100,000 tonnes annually. Slovenia is the subregional leader in wood chip consumption in district heating systems. In Serbia, the wood-based panels and juice industries are the largest consumers of wood chips (the juice industry uses wood chips as an alternative to heating oil). Ongoing projects in several heating plants will affect wood chip consumption in district heating systems in Serbia, but not before the heating season of 2017-2018 (Business Annual Report of District Heating in Sarajevo, Banja Luka, Prijedor, Sokkolac and Gradiska, 2015). Bosnia and Herzegovina's first private district heating system has been constructed, with wood pellet consumption in the heating season estimated at 30,000 tonnes. The estimated price of generating 1 MWh of heat delivered to households using wood pellets is about 18% less than the cost of equivalent heat using gas (Business Annual Report of District Heating in Sarajevo, Banja Luka, Prijedor, Sokkolac and Gradiska, 2015).

Woodfuel production in the western Balkans increased in 2015, to 214.2 PJ. Firewood (84% of total wood energy), wood pellets (7%) and wood chips (6%) are the main woodfuels used (graph 9.2.2). Wood energy supplies about 30% of the total energy produced in Bosnia and Herzegovina and Serbia and 16% of the total energy produced in Croatia. The share of wood energy ranges from 4% to 7% in other countries in the subregion (Glavonjić, 2016).

GRAPH 9.2.2

Contribution of woodfuels to wood energy production in the western Balkans, 2015

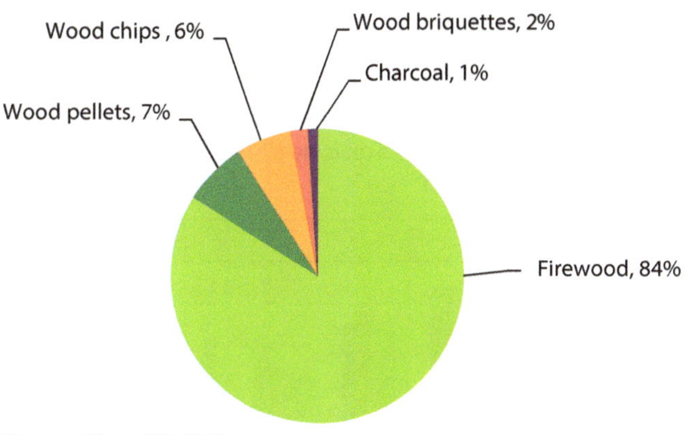

Source: Glavonjić, 2016.

9.2.2 Prices

Argus Media (2016a) reported that CIF spot prices for industrial wood pellets at Amsterdam, Rotterdam and Antwerp (ARA) declined steadily in 2015 through the first quarter of 2016 (graph 9.2.3). ARA CIF spot prices for industrial pellets declined significantly in 2015, with the largest fall (about 18%, year-on-year) reported in March 2016. This decline in prices can be associated with factors such as flat demand (because buyers had ample supplies of pellets in storage) and an increase in supply from European suppliers and from North American and Russian exporters. The residential (premium) wood pellet market exhibited a slight downward price trend: according to Argus Media (2016a), the price of delivered bulk premium (EN plus certified A1) pellets in northern Italy dropped by 2% in April 2016, to €145 per tonne. On the other hand, delivered bagged pellets showed almost no change, with an average price of €190 per tonne.

GRAPH 9.2.3

Wood pellet prices at Amsterdam, Rotterdam and Antwerp, May 2014-March 2016

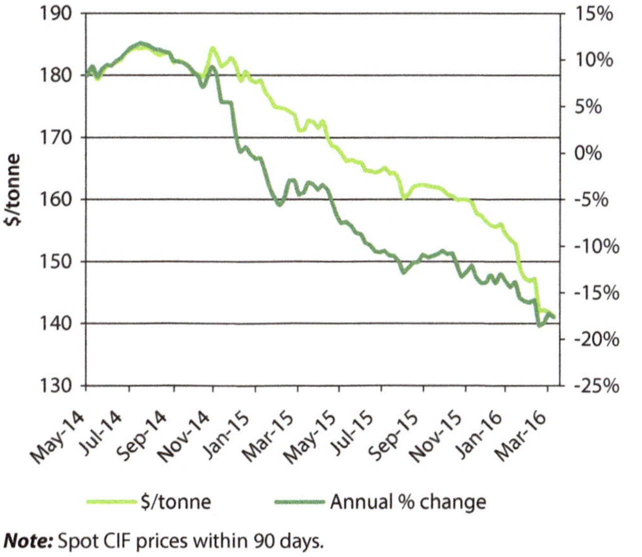

Note: Spot CIF prices within 90 days.
Source: Argus Media, 2016a.

Woodfuel pricing in the western Balkans is characterized by large country-level variation. For firewood, for example, Albania had the lowest prices in 2011-2015 and Croatia the highest, with a price differential of more than 100% per stacked m³ between the two countries (graph 9.2.4).

GRAPH 9.2.4

Average market price per stacked cubic metre of firewood in western Balkan countries, 2011-2015

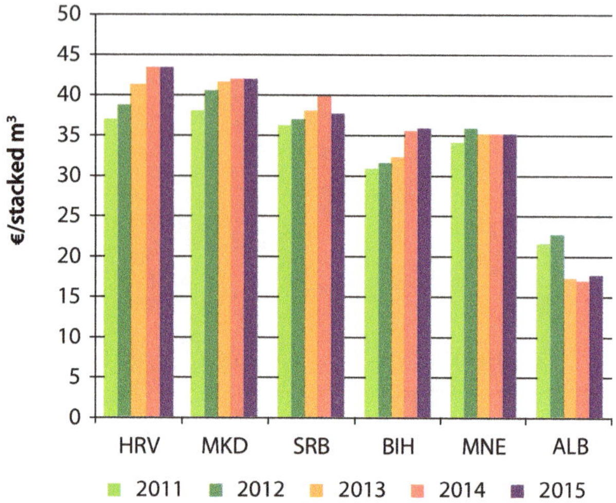

Note: HRV = Croatia; MKD = The former Yugoslav Republic of Macedonia; SRB = Serbia; BIH = Bosnia and Herzegovina; MNE = Montenegro; ALB = Albania. VAT included.
Source: Glavonjić, 2016.

Wood pellet prices in the western Balkans have experienced an overall upward trend, with a declining trend observed only in Slovenia (where, however, prices are, on average, still the highest among western Balkan countries) (graph 9.2.5). This is likely the result of market factors such as changing demand in the Italian market; moreover, the ample pellet stocks remaining at the end of 2014 led many pellet manufacturers to lower their prices in early 2015 to reduce stocks and improve cash flows.

GRAPH 9.2.5

Average market price per tonne of wood pellets in western Balkan countries, 2011-2015

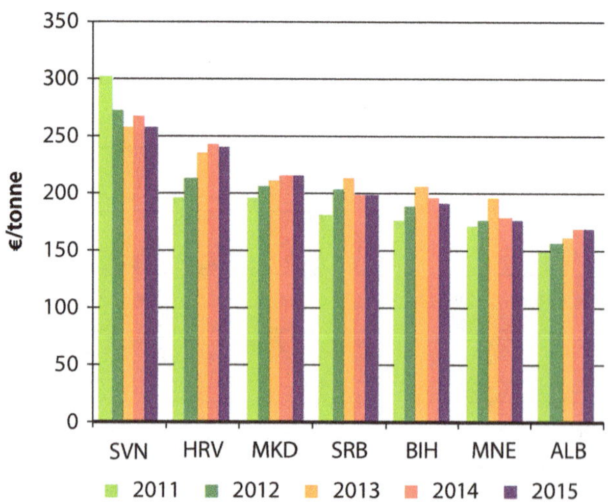

Note: SVN = Slovenia; HRV = Croatia; MKD = The former Yugoslav Republic of Macedonia; SRB = Serbia; BIH = Bosnia and Herzegovina; MNE = Montenegro; ALB = Albania. VAT included.
Source: Glavonjić, 2016.

9.2.3 Trade

The importance of imported feedstock to the overall production of energy from solid biofuels in the EU28 continues to grow (Eurostat, 2016b). Imported solid biofuels generated 310.4 PJ of energy in the EU28 in 2014, which was 9% of all solid biofuels used for primary energy production in the subregion in that year (graph 9.2.1). The EU28 imported an estimated 7.2 million tonnes of wood pellets in 2015; the US accounted for 60% of this, followed by Canada (21%) and the Russian Federation (11%) (graph 9.2.6).

Exports of wood energy from the western Balkans increased by 5.8% (by energy content) in 2015 compared with 2014, to 38 PJ. Firewood accounted for about 45% of the total, by energy content, followed by wood pellets (27%) and wood chips (18%); these values correspond to volumes of 1.88 million m³ of firewood, 679,000 tonnes of wood chips, and 583,000 tonnes of wood pellets. The export volume of firewood and wood pellets increased in 2015, but the volume of wood chips dropped slightly. The increase in wood pellet exports was due largely to an increase in pellet exports from Croatia (which accounted for 34% of all pellet exports from the subregion). Exports also increased from Montenegro and Slovenia, but they declined from Bosnia and Herzegovina and Serbia.

GRAPH 9.2.6

EU28 imports of wood pellets, 2010-2015

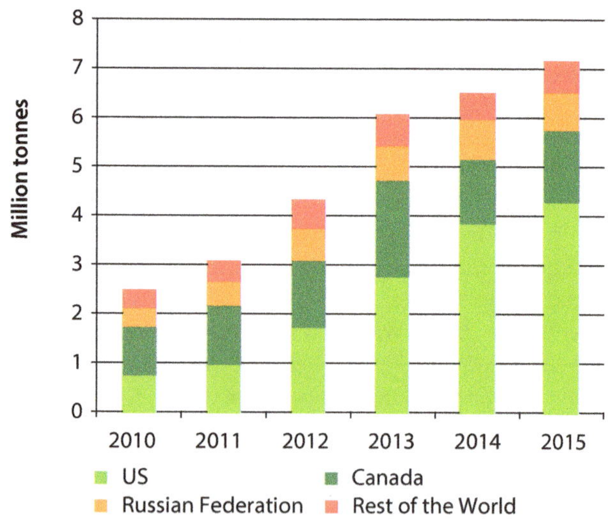

Source: Eurostat, 2016b.

An analysis of woodfuel trade flows among western Balkan countries showed that about 19% of the total trade occurred within the subregion in 2015, while 64% of total wood pellet production, 84% of total wood-briquette production, and 59% of total wood chip production were exported from the subregion. Italy remains the most important market for the export of all woodfuels from the western Balkans; about 71% of the total export of wood pellets went to the Italian market in 2015, which was 25% of Italy's total imports. About one-third of Italian firewood imports came from western Balkan countries (figure 9.2.1).

FIGURE 9.2.1
Wood-splitting and packing for export, Bosnia and Herzegovina

Note: Firewood exported to Italy mainly consists of split logs 33 cm long, packed in pallets (1.8 x 1 x 1 m), with 20-25% moisture content.
Source: Glavonjić, 2016.

9.3 COMMONWEALTH OF INDEPENDENT STATES

9.3.1 Consumption and production

Demand for wood energy feedstock is increasing in the CIS as wood energy consumption (including pellets, briquettes and chips) grows in the subregion and neighbouring countries. Wood energy production increased in part due to the devaluation of the rouble (by 68% in 2015-2016), which also affected the currencies of other CIS states with close economic ties (for many post-Soviet states, trade with the Russian Federation represents more than 5% of their GDP). The devaluation of national currencies supported an increase in export sales of wood products (including wood energy products): production costs stayed relatively unchanged but the value of sales nearly doubled. Tied to changes in exchange rates, wood pellet manufacturing for export enjoyed very large increases in revenue in 2015. Wood pellet manufacturers have reportedly allocated some of this increased revenue to support capital investments.

Most district heating plants in the CIS are old and inefficient, and there is strong interest among local governments in modernizing heating plants and reducing wood energy costs. For example, many boiler houses in the Arkhangelsk oblast have already switched from coal or oil to wood pellets and chips, driven by production cost savings – the cost of locally available woodfuels (per unit of generated energy) in the subregion is 23% less than that for coal and oil (with the latter unavailable locally). In 2015, the local government in the Arkhangelsk area subsidized the collection of wood that otherwise would have been disposed of to increase use and reduce illegal dumping. The positive experience in Arkhangelsk was reproduced in the Komi Republic, where in 2015 the local government covered the cost of planning and constructing waste-wood collection areas, while the cost of producing wood pellets and fuel briquettes was subsidized directly from the regional budget. Financial support was provided in the form of direct subsidies for some companies and through tax deductions for woodfuel producers, and state subsidies gave consumers additional support for using wood energy. Woodfuels are cost-competitive with coal and oil in the Komi Republic within 100 km of railway lines.

Fuel-switching to wood feedstock occurred in several regions in the CIS in 2015. For example, coal was unavailable for technical and logistical reasons in parts of northwest Russian Federation for several days of extremely cold winter weather, and boiler houses quickly switched to fuel briquettes to avoid the freezing of district heating networks.

According to the official Russian statistics agency, Rosstat, total wood pellet production increased by 6.5% in the Russian Federation in 2015 compared with 2014, to 973,000 tonnes, the country's highest production since 2009. Pellet production increased throughout 2015, except in September-November, when it decreased. The majority of wood pellets were produced in the northwest, although production there was lower than in 2014. The country's new pellet-producing areas, the Far East and Central regions, experienced rapid production growth from both new and existing plants, driving up overall output. There are plans to establish new pellet factories in the Habarovsk region, the Vologra region, the Irkutsk region and others, with the biggest factories planned in the Irkutsk region (including two factories with a capacity of 105,000 tonnes and 75,000 tonnes, respectively).

9.3.2 Prices

Domestic wood pellet prices increased in the Russian Federation in 2015 (graph 9.3.1). The rouble-denominated cost of wood energy products rose due to higher production costs and increased competition. Unfavourable weather conditions, especially in winter, are generally associated with price peaks and other fluctuations. US dollar-denominated prices of wood pellets exported from the Russian Federation to Denmark, Finland, Germany, Latvia and the Republic of Korea decreased in late 2015 and early 2016 and the export volume increased. As of the end of 2015, export FOB prices for pellets were fluctuating between US$100 and US$117 per tonne.

GRAPH 9.3.1

Domestic prices for wood pellets in the Russian Federation, January 2014-February 2016

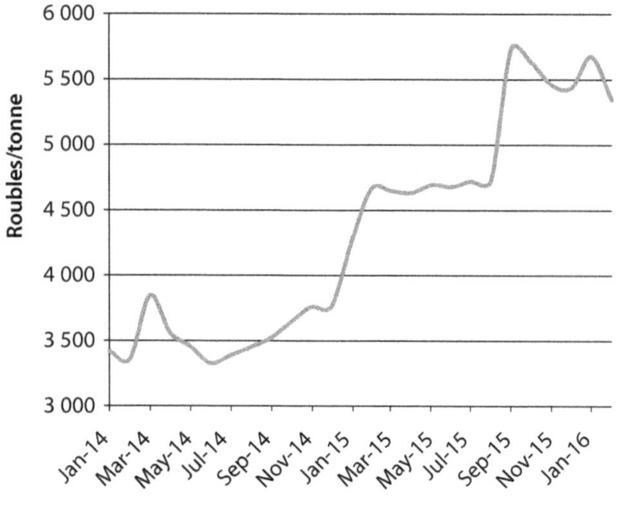

Source: Russian Timber Journal, 7/8-2016.

9.3.3 Trade

Net exports of wood pellets (i.e. exports net of imports and re-imports) from the Russian Federation increased by 6% in 2015 compared with 2014, to 932,700 tonnes (COMTRADE, 2016), but the trade value dropped by 20% when accounted for in US dollars (due to the depreciation of the rouble). The Leningrad region exported the most wood pellets (via the harbour), followed by the Republic of Karelia (the majority of exports going to Finland) and then Saint Petersburg. The biggest growth in pellet exports in 2015 was in the Irkutsk region, where there was a 710-fold increase. The biggest reduction in exports was in the Republic of Mari El, where there was a six-fold decrease.

Europe is the main destination for wood pellets manufactured in the Russian Federation. The biggest export market in 2015 was Denmark, which took 381,000 tonnes, followed by Sweden (154,000 tonnes), Germany (73,000 tonnes) and the Republic of Korea (72,000 tonnes).

It is expected that demand for wood energy feedstock (especially wood chips) in the Russian Federation will increase further due to positive export prospects. For example, Finland plans to open new renewable energy facilities that will consume woodfuel, and the decision of several energy-generating companies to switch from fossil fuels to wood chips and wood pellets could further increase trade. The ongoing geopolitical crisis in Ukraine has heightened the perceived commercial risks, and this was reported to be a major obstacle for Finnish companies trading with suppliers in the Russian Federation (Luke, 2014). Nevertheless, Karvinen and Mutanen (2015) reported that, as of the beginning of 2015, such geopolitical tensions had not had a direct, discernible influence on the wood trade between Finland and the Russian Federation.

9.4 NORTH AMERICA

9.4.1 Consumption and production

There was a major increase in the consumption of wood pellets in Canada in the ten years to 2015 (graph 9.4.1). The recent spike might have been driven by colder-than-historical-average winters in eastern Canada in 2013-14 and 2014-15, with the consumption of wood pellets reaching an estimated 600,000 tonnes in 2015, which was about 27% of the country's total supply (Wood Pellet Association of Canada, 2016). Canada's installed bio-based electricity generation capacity is now over 2 GW, most of it in British Columbia (827 MW of capacity) and Ontario (681 MW) (Natural Resources Canada, 2016a); at conservative capacity factors of 50%, this translates to approximately 31 PJ of bio-based electricity. Canada's consumption of wood energy from residues continues to rise, reaching 261 PJ in 2014, 6% higher than in 2010. Canada also produced about 247 PJ from spent pulping liquor in pulp and paper mills in 2014, a 6% rise over 2010 despite a drop in chemical pulp outputs (Statistics Canada, 2016b). Increases in spent pulping liquor consumption is driven in part by the Pulp and Paper Green Transformation Program, which invested CAD 1 billion to upgrade generation capacity, ultimately supporting 98 projects across the country (Natural Resources Canada, 2016b). Biomass now generates approximately 381 PJ of industrial energy in Canada, which is almost 18% of the total energy requirement of the country's manufacturing sector (Statistics Canada, 2016c). The residential sector in Canada is estimated to be consuming 171 PJ of wood energy for heating (Statistics Canada, 2016b). Wood pellets contribute only a fraction – 4-8 PJ – of this wood-based residential heating, consuming from 250,000 to 500,000 tonnes of pellets annually (Rebiere, 2016; Wood Pellet Association of Canada, 2016). Wood energy for heating is concentrated in the provinces of Ontario and Quebec, which together account for some 60% of the national residential consumption of wood energy for heat (Rebiere, 2016). This high consumption is driven largely by population size, although Quebec has promoted biomass-based heat. Projects relying on woodfuels to generate electricity (e.g. the Thunder Bay and Atikokan generating stations in Ontario) have the potential to increase consumption, but they continue to face challenges of commercial feasibility. The Atikokan Generating Station reportedly could consume up to 90,000 tonnes of pellets per year (Walters, 2015).

GRAPH 9.4.1

Annual consumption and production of wood pellets in Canada, 2000-2015

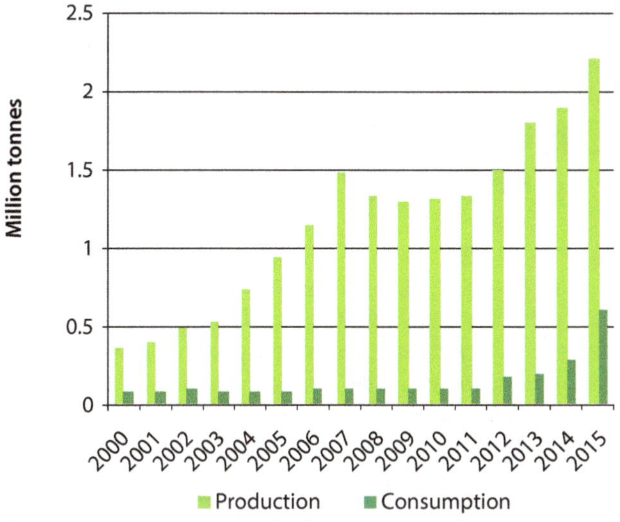

Sources: FAOSTAT, 2016; Wood Pellet Association of Canada, 2016a; Statistics Canada, 2016a.

Canada is the world's third-largest wood pellet producer after the US and Germany, providing slightly less than 10% of global production (FAOSTAT, 2016). Canada has 37 operational pellet-manufacturing plants with an estimated total installed capacity of about 4 million tonnes (Rebiere, 2016). Canadian production of wood pellets continues to lag behind capacity, partly because of persistent fibre shortages in eastern Canada (Macklin, 2016). At least two facilities, with a combined capacity of 200,000 tonnes per year, were taken offline in 2015: Viridis Energy ceased operations at the Okanagan Pellet Co. plant in Kelowna, British Columbia, due to high costs (Bioenergy-news.com, 2016), and Boreal Pellet in Amos, Quebec, closed after a fire (Macklin, 2016). Thirteen facilities are in the planning or development stages (Canadian Biomass Magazine, 2016).

Summary statistics from the Monthly Energy Review (US Department of Energy, 2016a) indicate that national wood consumption in the US was 2,173 PJ in 2015 (graph 9.4.2), a decrease of about 7.6% compared with 2014. Most of the decline can be attributed to a reduction in residential wood energy consumption of 140 PJ, which in turn can be linked to the relatively mild winter and low-priced fossil fuels. Wood comprised 23% of all renewable energy consumption in the US in 2015, down from 25% in 2014 (US Department of Energy, 2016a). Of the wood-energy-consuming sectors, the industrial sector accounts for the largest share, at about 63%, followed by the residential sector, at 22%. The consumption of these two sectors has fluctuated over time, and the sector with the largest growth is electricity generation.

The consumption of wood energy for electricity generation in the US increased from 133 PJ in 2000 to 260 PJ in 2015. In contrast, however, per capita wood use for energy declined from 1990 to 2001, after which it levelled off at about 0.8 m^3 per year (US Department of Energy, 2016a; US Bureau of Census, 2016). Among the factors associated with lower per capita wood energy consumption are lower prices for alternative energy sources for residential heating, urbanization, and income (Song et al., 2012). Overall, the likely short-term outlook is for little change in the domestic consumption of wood energy (US Department of Energy, 2016b), in part because small reductions in industrial use are being offset by small increases in other sectors (e.g. power generation).

There were 107 operational wood pellet manufacturing facilities in the US at the beginning of 2016 (Forisk, 2016), with an estimated combined installed capacity of 10.3 million tonnes; total wood pellet production was 6.9 million tonnes in 2014 (FAOSTAT, 2016). A comparison of estimated capacity and production suggests that wood pellet plants were operating at about 80% of installed capacity in 2014. Large manufacturers of industrial pellets are actively seeking third-party certification to address sustainability concerns and account for carbon emissions. For example, all Enviva-owned pellet plants are now certified to the Sustainable Forestry Initiative (SFI) fibre sourcing standard and by the Green Gold Label programme – the latter covering the chain of custody, processing, and GHG and energy balance accounting. Enviva-owned plants also have chain-of-custody certification with the Forest Stewardship Council and the SFI/Programme for the Endorsement of Forest Certification (PEFC) (Enviva, 2016). Large wood pellet manufacturers and users state that most of the feedstock for the manufacture of pellets comes from low-grade fibres sourced from logging and mill residues (American Wood Council, 2015). Nevertheless, surveys of existing and proposed pellet production facilities indicate that mill residues are expected to comprise only 25% of feedstock for pellets, and logging residues and waste products are not cost-competitive at current prices. More information on feedstock should be forthcoming soon from the US Energy Information Administration's Densified Biomass Fuel Report (EIA-63C), a mandatory survey of manufacturers of densified biomass fuel products in the US, which requires pellet producers to identify their wood sources, including species for roundwood (by size and quality), wood chips from chip mills, logging residues, mill residues, bark, post-consumer wood products, and dedicated energy crops. EIA-63C respondents were notified in December 2015, and data collection was initiated in February 2016 (Pellet Wire, 2016).

GRAPH 9.4.2

US annual wood energy consumption by sector, and corresponding per capita wood consumption, 1990-2015

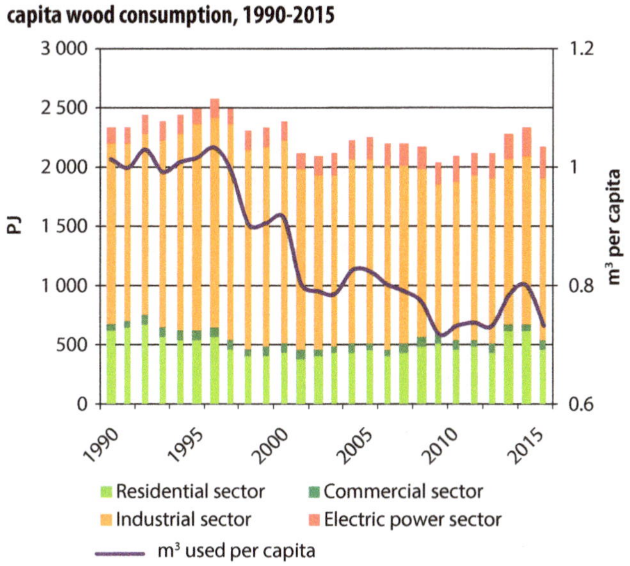

Source: US Department of Energy, 2016a.

9.4.2 Prices

Canadian wood pellets could be purchased on the retail market for CAD 220-250 per tonne in spring 2016 (Gildale Farms, 2016; PelletStoveStore, 2016), compared with the average price on the international market of close to CAD 175 per tonne FOB (Statistics Canada, 2016a). Canadian prices were slightly (about 5%) higher in spring 2016 than in 2014, year-on-year. The value of wood chips was also higher, approaching CAD 95 per tonne in 2015 compared with CAD 80 per tonne in 2014 (Statistics Canada, 2016a). Higher prices for bioenergy commodities may not be good news for the industry, however, given the slump in fossil fuel commodity pricing – it suggests that Canadian bioenergy products may be less competitive in the global marketplace.

In the US, domestic wood pellet retail prices (excluding delivery) reported by the State of Massachusetts Office of Energy and Environmental Affairs (2016) in winter 2016 were $260 per tonne (bulk) and $6.73 per 18.1 kg (40 lb) bag. In the US Northeast region, the price of premium wood pellets ranged from $250 to $290 per tonne in spring 2016, and the price of super premium pellets[23] ranged from $308 to $330 per tonne (BT Enterprises, 2016). These were all slight declines compared with the same period in 2015, with the drop in prices likely associated with weaker seasonal demand due to milder weather. Market uncertainty associated with potential changes to renewable energy targets in the EU28 seems to have led to lower delivered prices for pellets, affecting suppliers to European industrial consumers (graph 9.2.3). Argus Media (2016a) reported that "with both the premium and industrial markets grappling with high levels of uncertainty over how much demand is expected to come online in the next two years, many market participants are eyeing the market with increasing caution and are hesitant to consider trading beyond the first quarter of 2017".

9.4.3 Trade

North America exported 6.2 million tonnes of wood pellets in 2015 (after accounting for re-exports of Canadian pellets via the US), with 1.6 million tonnes and 4.6 million tonnes reported by Canada and the US, respectively (COMTRADE, 2016). Canadian wood pellet exports in 2015, valued at CAD 284 million, were steady compared with 2014 and slightly down on 2013 values (Statistics Canada, 2016a). The top destination for Canadian wood pellet exports in 2015 was the UK (74%), followed by the US (13%) and Italy and Japan (5% each). This distribution was similar to 2014, but it is clear that the reliance of Canadian pellet exporters on the UK market has increased in recent years, likely due to ongoing purchase agreements between Rentech and Drax Power in the UK (McCormick, 2014).

The US exported nearly 4.6 million tonnes of wood pellets in 2015, which was a 13% increase over 2014. Previously, exports (by tonnage) had increased by 52% in 2013 compared with 2012 and by 41% in 2014 compared with 2013 (US International Trade Commission, 2016). The UK continues to be the largest importer of US pellets, accounting for about 84% of the tonnage and value traded in 2015. This was an increase compared with 2014, when the UK was the destination for some 73% of the tonnage of US exported pellets.

9.5 POLICY, STANDARDS AND REGULATORY INFLUENCES

The UNFCCC's Paris Agreement has created a set of policy goals with potential to influence the role that wood energy plays in national energy portfolios. Each signatory to the Paris Agreement is expected to develop nationally determined contributions to reducing GHG emissions, and the role of wood energy is therefore likely to vary from country to country (United Nations Treaty Collection 2015). For example, Canada has stated a national long-term goal of decarbonizing its economy and shifting from fossil fuels (Canada 2015). The most crucial policy tool under discussion at the federal level for achieving this aim is a national price on carbon (Cheadle, 2016). The adoption of a federal policy on this would mirror action taken by four of ten provinces. An economy-wide carbon tax has been in place in British Columbia since 2008, currently set at CAD 30 per tonne (Government of British Columbia, 2016); the Government of Alberta recently pledged to impose a carbon levy that would reach CAD 30 per tonne by 2018 (Government of Alberta, 2016). Quebec has a cap-and-trade system that is pricing carbon at about CAD 16 per tonne (Government of Quebec, 2016), and Ontario is moving towards implementing a cap-and-trade system in 2017 that would likely see prices similar to those in Quebec (Government of Ontario, 2016). The impact of carbon prices on the bioenergy market is unclear, with some researchers expecting they will increase bioenergy uptake (Peterka, 2015). In practice, jurisdictions like British Columbia have not seen major increases in bioenergy use under a carbon tax regime, perhaps because of the "revenue-neutral" design of the tax system, which offsets carbon bills with lower income or corporate tax (Bradburn, 2014).

As the backbone of the US commitment to the Paris Agreement, the US Environmental Protection Agency (USEPA) has outlined GHG emissions reduction targets for power plants. Targets and tools for achieving reductions are outlined in the US Clean Power Plan, which could potentially encourage the greater use of wood for energy. The extent to which the Plan affects wood as a feedstock in power generation will hinge on the prices of other renewable energy sources and the biogenic carbon emissions guidelines of the USEPA (McCabe, 2015). However, the US Supreme Court has stayed the implementation of the Clean Power Plan, pending a judicial review (USEPA, 2016). Concerns that the Plan could curtail the potential carbon-reducing benefits of wood energy have led to a filing for relief from wood suppliers (American Wood Council, 2015). Legislation being discussed in the US Congress supports the role of US forests in addressing national energy needs; US Senate Bill 114-2012 is

[23] Super premium pellets are intended for applications where a very low ash content (less than 0.5%) is desirable.

directed at the departments of Energy and Agriculture, along with the USEPA, and has two aims: 1) to ensure a federal policy that is consistent across federal government agencies and recognizes the energy, conservation and forest management benefits of using forest biomass; and 2) to establish clear, simple policies for the use of forest biomass as an energy solution, including policies that reflect the carbon-neutrality of forest bioenergy – provided that the use of forest biomass for energy production does not cause the conversion of forests to non-forest use, encourages private investment throughout the forest biomass supply chain, and improves forest health.

In Europe, the Dutch government has released draft sustainability criteria and means of compliance for solid biomass for open consultation under the SDE+ ("Encouraging Sustainable Energy Production"). The SDE+ is an operating grant for producers to provide financial compensation for selected unprofitable (compared with fossil fuels) renewable energy generation. The SDE+ is available for the production of renewable electricity, renewable gas, renewable heat and CHP. Grant recipients can demonstrate compliance with the sustainability criteria via certification (e.g. FSC, PEFC, SFI, American Tree Farm System, and Canadian Standards Association); if none is available, verification can be used to demonstrate compliance (a verification protocol is under development). The current draft identifies five types of solid biomass, of which three are forest-based: 1) woody biomass from large forest management units (≥500 hectares); 2) woody biomass from small forest management units (<500 hectares); and 3) residues from nature and landscape management. All three categories will have to meet sustainable management criteria and chain-of-custody principles. Certification/verification of carbon and land-use change are only required for woody biomass obtained from forest management units. Following consultation with stakeholders, a final protocol is expected to be published in September 2016 (Netherlands Enterprise Agency, 2016). The UK has announced a reduction of subsidies and tax benefits (which were financed through utility bills) for generating renewable energy from biomass (UK Department of Energy and Climate Change, 2015). There are a number of reasons for the reduction – including advances in the efficiency of renewable energy technology, higher-than-expected demand-led feed-in tariffs, and the Renewables Obligation – which, collectively, are expected to result in the generation of more renewable electricity than previously projected. The EU has commissioned a study on the impacts of pellet production on southern US forests to help in setting targets for solid fuels and their role in meeting renewable energy aims beyond 2020 (Tenders Electronic Daily, 2015).

Public policies and programmes affecting forests and wood energy continue to evolve in the western Balkans. The Albanian parliament has approved a ten-year moratorium on logging for industrial purposes and export (due to the deteriorating condition of Albanian forests). The law has a provision allowing local authorities to cut a limited amount of wood for heating (IHB, 2016).

Source: D. Hanlon, 2016.

As illustrated by the Sustainable Biomass Partnership (SBP) and ENplus certification, private-sector stakeholders continue to develop and adopt standards for wood pellet sustainability and quality. The SBP has introduced a new funding model, in which SBP certificate-holders and any entity taking legal ownership of SBP-certified biomass and selling supply with an SBP claim will pay fees to support the SBP; previously, only utilities supporting the scheme had paid a membership fee. The fees for wood pellet producers will be €0.15 ($0.17) per tonne sold, effective from 1 October 2016, and wood chip producers will pay €0.08 per tonne, effective from 1 April 2017. Traders will be charged an annual fee, depending on volume: those handling more than 250,000 tonnes per year will pay €25,000, those trading between 100,000 tonnes and 249,900 tonnes per year will pay €10,000, and those selling less than 100,000 tonnes per year will pay no fee (Argus Media, 2016a). The SBP has approved two certification bodies, NEPCon and NSF International, to certify compliance with SBP standards. In the US, the first SBP certificate-holders include Georgia Biomass and Varn Wood Products; in both cases, the certificates were issued in 2016 (Sustainable Biomass Partnership, 2016). ENplus certification is being adopted increasingly widely: some 8 million tonnes of ENplus certified pellets were expected to be produced globally by the end of 2015 (Calderon et al., 2016).

9.6 INNOVATION

Small-scale, highly-efficient CHP units using woodfuels present opportunities for development. Some seem close to commercial market feasibility, although financial hurdles remain. An example is Entrade Energiesysteme's E3 biomass CHP plant for decentralized energy supply (Institution of Mechanical Engineers, 2015): it has a high-temperature reactor to generate syngas (synthesis gas) from solid biomass, which is then burned to generate electricity. E3 units can reportedly be produced at €2,500 per kW installed capacity, which is price-competitive with large-scale biomass power plants; in the UK, the E3 is able to deliver electricity at 6.3 pence per kWh. A single E3 system can generate a total of 8 GWh of electricity and over 19 GWh of heat per year.

There has been little progress, however, in the production of advanced biomass-based liquid fuels at commercial and research scales. The development of such fuels is highly dependent on policy targets, which seem to be at a cross-point in both the EU and the US. In the EU, the primary production of biofuels was 12,800 tonnes of oil equivalent in 2013. The Renewable Fuels Standard Guidelines published by the USEPA in November 2015 indicate that cellulosic ethanol production was about 9.5 million litres in 2015, with a potential range of production in 2016 of 0 to 83 million litres, indicating considerable uncertainty about the production of this fuel (USEPA, 2015). This production occurred in only two commercially operating cellulosic ethanol plants -- neither of which was using wood. In Canada, the Enerkem project in Edmonton, Alberta, has begun production of methanol and plans to produce ethanol from municipal solid waste. This project, although delayed from its original timeline, represents a technologically new approach to renewable fuel production (Enerkem, 2016).

The use of torrefied wood pellets could reach commercial scale in Canada. The Thunder Bay Generating Station in Ontario, a converted coal-fired power plant, has been built to operate on torrefied biomass, although there are no local suppliers of this fuel in Ontario. Should the plant prove successful it will provide a template for the conversion of other coal-fired plants, but questions remain about its cost-effectiveness. According to Ontario Power Generation (2016), the Thunder Bay Generating Station project has received public attention because of the relatively low cost of conversion (CAD 5 million, compared with CAD 170 million for the Atikokan power facility using pellets), but energy generation costs seem much higher. Walters (2015) noted that power generation costs at the Thunder Bay Generating Station were about 25 times those of other biomass plants in Ontario, due largely to the lack of local suppliers of, or a competitive market for, torrefied pellets.

9.7 REFERENCES

American Wood Council. 2015. Paper, wood products manufacturers file biomass litigation on clean power plan. Available at: http://awc.org/news/2015/12/22/paper-wood-products-manufacturers-file-biomass-litigation-on-clean-power-plan

Argus Media. 2016a. Biomass markets. Issue 16-014. Available at: www.argusmedia.com/~/media/files/pdfs/samples/argus-biomass.pdf/?la=en

Argus Media. 2016b. German Pellets administrator agrees to sell plants. Available at: www.argusmedia.com/news/article/?id=1232957

Bioenergy-news.com. 2016. Viridis Energy ceases operations at its Okanagan Pellet Co. plant. Available at: www.bioenergy-news.com/display_news/10372/

Bradburn, K. 2014. 2014 Canbio report on the status of bioenergy in Canada. Canadian Bioenergy Association. Available at: www.fpac.ca/wp-content/uploads/2014_CanBio_Report.pdf

BT Enterprises. 2016. Available at Price List. http://btpellet.com/page/price-list

Business Annual Report of District Heating Systems in Sarajevo, Banja Luka, Prijedor, Sokolac and Gradiška. 2015. Internal documents.

Calderon, C., Gauthier, G. and Jossart, J. 2016. AEBIOM statistical report 2015. European bioenergy outlook. 206 pp.

Canada, 2015. Canada's National Statement at COP21 - Address by the Right Honourable Justin Trudeau. Ottawa, Canada: Prime Minister of Canada. Available at http://pm.gc.ca/eng/news/2015/11/30/canadas-national-statement-cop21

Canadian Biomass Magazine. 2016. Pellet Map 2015. Available at: http://www.canadianbiomassmagazine.ca/images/stories/pellet-map2015.pdf

Cheadle B. 2016. Premiers agree to carbon pricing as part of national climate strategy. National Post, March 3, 2016.

COMTRADE. 2016. United Nations Commodity Trade Statistics Database. HS 440131. Statistics Division. New York. Available at: http://comtrade.un.org/data/

Enerkem 2016. Biofuels and green chemicals. Available at: enerkem.com

Enviva. 2016. Certifications. Available at: www.envivabiomass.com/sustainability/certifications

Eurostat. 2016a. Energy from renewable sources. Available at: http://ec.europa.eu/eurostat/statistics-explained/index.php/Energy_from_renewable_sources

Eurostat. 2016b. International trade, DS-045409: EU trade since 1988 by HS2, 4, 6 and CN8. Includes HS 4401.30.20 (2009-2011) and HS 4401.31 (2012-2014).

Eurostat. 2016c. Supply, transformation and consumption of renewable energies: annual data [nrg 107a]. Available at: http://ec.europa.eu/eurostat/web/energy/data/database

FAOSTAT. 2016. Forestry statistics. Rome. Available at: http://faostat.fao.org/site/626/DesktopDefault.aspx?PageID=626#ancor

Forisk. 2016. Wood Bioenergy US. Available at: http://forisk.com/resources/resources-from-forisk-wood-bioenergy-us-free-summary/

Gildale Farms. 2016. Fuel Pellet prices. Available at: http://www.gildalefarms.ca/fuel-pellets/

Glavonjić, B. 2016. Personal communication. Data from the University of Belgrade Timber Trade Center database. Belgrade, March 2016.

Government of Alberta. 2016. Carbon levy and rebates. Edmonton, Canada. Available at: www.alberta.ca/climate-carbon-pricing.cfm

Government of British Columbia. 2016. Carbon tax. Victoria, Canada, British Columbia Ministry of Finance. Available at: www.fin.gov.bc.ca/tbs/tp/climate/carbon_tax.htm

Government of Ontario. 2016. Cap and trade. Toronto, Canada. Available at www.ontario.ca/page/cap-and-trade

Government of Quebec. 2016. The Quebec cap and trade system for greenhouse gas emissions allowances. Quebec City, Canada. Available at: www.mddelcc.gouv.qc.ca/changements/carbone/Systeme-plafonnement-droits-GES-en.htm

IHB. 2016. The Albanian parliament has approved a 10 year moratorium on logging for industrial purposes and export. Available at: www.ihb.de/wood/news/Albania_logging_ban_45823.html

Institution of Mechanical Engineers. 2015. World's smallest biomass plant reaches operational landmark. Available at: www.imeche.org/news/engineering/world%27s-smallest-biomass-plant-reaches-operational-landmark-20071503

Karvinen, S. and Mutanen, A. 2015. Puumarkkinoiden toimivuus Luoteis-Venäjällä yritysten näkökulmasta [Functioning of roundwood markets in Northwest Russia - company view]. Unpublished manuscript.

Luke 2014. Uutiset: Venäjällä on suunnitteilla lisämaksu yli 12 tonnin kuljetuksille kaikille teille (News: Russia Federation plans additional fee for transports over 12 tonnes). 2014. Idän metsätieto-palvelun uutinen 19.12. Available at: http://www.idanmetsatieto.info/fi/?ID=270&news=view&newsID=2803

Macklin, A. 2016. State of the industry: analyzing the health of biomass markets entering 2016. Canadian Biomass, January/February: 14-17.

McCabe, J. 2015. The role of biomass in achieving clean power plan goals: a 2016 workshop to foster a constructive discussion. US Environmental Protection Agency. Available at: https://blog.epa.gov/blog/2015/11/the-role-of-biomass

McCormick C. 2014. Export-worthy: a robust supply chain was critical for attracting a major European wood pellet client. Canadian Shipper. Available at: www.canadianshipper.com/features/export-worthy-a-robust-supply-chain-was-critical-for-attracting-a-major-european-wood-pellet-client/

Netherlands Enterprise Agency. 2016. SDE+ sustainability requirements for solid biomass. Available at: http://english.rvo.nl/sites/default/files/2016/03/SDE%20Sustainability%20requirements%20for%20solid%20biomass.pdf

Natural Resources Canada. 2016a. About renewable energy. Ottawa. Available at: www.nrcan.gc.ca/energy/renewable-electricity/7295

Natural Resources Canada. 2016b. Pulp and paper green transformation program: mission accomplished. Ottawa. Available at: www.nrcan.gc.ca/forests/federal-programs/13141

Ontario Power Generation. 2016. Thunder Bay Generating Station. Website. Available at: www.opg.com/

generating-power/thermal/stations/thunder-bay-station/Pages/thunder-bay-station.aspx

PelletStoveStore. 2016. Wood pellets. Website. Available at: https://pelletstove.ca/wood_pellets.htm

Pellet Wire. 2016. EIA data collection underway. Available at: www.pelletheat.org/pelletwire-2-05-2016

Peterka, A. 2015. Climate: imposing carbon price would spur bioenergy, slash emissions – MIT study. Energywire. Available at: www.eenews.net/stories/1060011746

Rebiere, M. 2016. Canadian market update. European Pellet Conference. Wels, Austria.

Song, N., Aguilar, F.X., Shifley, S. and Goerndt, M. 2012. Factors affecting wood energy consumption by U.S. households. Energy Economics, 34(2): 389–397.

Statistics Canada 2016a. Canadian International Merchandise Trade Database. Ottawa. Available at: www5.statcan.gc.ca/cimt-cicm

Statistics Canada. 2016b. Consumption of solid wood waste and spent pulping liquor for energy production. Cansim table 128-0018. Ottawa. Available at: www5.statcan.gc.ca/cansim/a26?lang=eng&retrLang=eng&id=1280018&pattern=&csid=

Statistics Canada. 2016c. Energy fuel consumption of the manufacturing sector, by fuel type. North American Industry Classification System (NAICS), table 128-0006. Ottawa. Available at: www.statcan.gc.ca/tables-tableaux/sum-som/l01/cst01/prim74-eng.htm

State of Massachusetts Office of Energy and Environmental Affairs. 2016. Wood pellet price survey. Available at: http://www.mass.gov/eea/energy-utilities-clean-tech/home-heating/wood/wood-dealers-and-prices.html

Sustainable Biomass Partnership. 2016. Approvals and certifications. Available at: www.sustainablebiomasspartnership.org/approvals-and-certifications/certified-biomass-producers

Tenders Electronic Daily. 2015. Study on the environmental implications of the increased reliance of the EU on biomass for energy imported from North America. 2015/S 012-016042. Available at: http://ted.europa.eu/udl?uri=TED:NOTICE:16042-2015:TEXT:EN:HTML&tabId=1

UK Department of Energy and Climate Change. 2015. Controlling the cost of renewable energy. Press release, 22 July 2015. Available at: www.gov.uk/government/news/controlling-the-cost-of-renewable-energy

United Nations Treaty Collection. 2015. Chapter XXVII Environment. 7d Paris Agreement. Available at: https://treaties.un.org/pages/ViewDetails.aspx?src=TREATY&mtdsg_no=XXVII-7-d&chapter=27&lang=en

US Bureau of the Census. 2016. Population estimates, July 1, 1990 to July 1, 2015. Available at: www.census.gov/popest/data/intercensal/index.html

US Department of Energy. 2016a. Monthly energy review. Release date: 29 March 2016. Tables 10.2 a, b, and c. Available at: www.eia.gov/totalenergy/data/monthly/#renewable

US Department of Energy. 2016b. Short-term energy and summer fuels outlook: renewables and carbon dioxide emissions. Release date: 12 April 2016. Available at: www.eia.gov/forecasts/steo/report/renew_co2.cfm

USEPA. 2015. Final renewable fuel standards for 2014, 2015 and 2016, and the biomass-based diesel volume for 2017. US Environmental Protection Agency (USEPA). Available at: www.epa.gov/renewable-fuel-standard-program/final-renewable-fuel-standards-2014-2015-and-2016-and-biomass-based

USEPA. 2016. Clean power plan for existing power plants rule summary. US Environmental Protection Agency (USEPA). Available at: www.epa.gov/cleanpowerplan/clean-power-plan-existing-power-plants

US International Trade Commission. 2016. Dataweb query on US exports of wood pellets. Available at: https://dataweb.usitc.gov Walters, J. 2015. Ontario's Auditor General report slams use of Thunder Bay, Atikokan generating stations. Available at: www.cbc.ca/news/canada/thunder-bay/ontario-s-auditor-general-report-slams-use-of-thunder-bay-atikokan-generating-stations-1.3349523

Wood Pellet Association of Canada. 2016. Wood pellet production. Available at: www.pellet.org/production/2-production

10 VALUE-ADDED WOOD PRODUCTS

Authors: ***Chris Gaston, Tapani Pahkasalo***

HIGHLIGHTS

- Value-added wood products markets are recovering unevenly; US import markets have grown rapidly to reach pre-global financial crisis levels, but European markets are growing more slowly and are even stagnant for some products.
- Global furniture production was worth $410 billion in 2015 and the value of the furniture trade was an estimated $140 billion. Trade policies continue to influence the investment decisions of manufacturers.
- House remodelling activity in the US is picking up as pent-up demand finally materializes; the forecast is for up to 9% annual growth in the remodelling market.
- Furniture companies are increasingly looking for ways to shorten delivery times to customers. Customization and designer furniture collections are aimed at increasing the competitiveness of production in the UNECE region.
- In the US, higher consumer confidence and increasing house prices are encouraging homeowners to start large-scale home improvement projects.
- More than 50% of consumers in the US use mobile devices to access furniture stores online, up from only 5% in 2010; traditional furniture stores are under pressure to attract customers and adapt their business models to new ways of shopping.
- Profiled-wood markets are improving, and producers face a challenge in ramping up capacities. Adding production capacity is relatively easy, but rebuilding sales organization and distribution channels, especially for retail customers, takes more time and is unlikely to happen at the same scale as before, even if markets grow faster again.
- North American production of glulam, wooden I-beam and laminated veneer lumber has made consistent significant gains from 2010 through to the forecast for 2016, tied to increased new-home construction.
- Cross-laminated timber (CLT) production is expanding globally, with new production facilities outside the DACH countries – Germany, Austria and Switzerland – where production originally began. Global CLT production was estimated at 650,000-700,000 m^3 in 2015.
- In Europe, CLT is 10-15% more expensive than masonry or cement for construction, but it is hoped that costs will come down with the further development of the industry and the standardization of products.
- The use of CLT in buildings is gaining momentum quickly in North America, thanks in large part to the interest shown in west-coast cities with strong wood cultures (e.g. Portland, Seattle and Vancouver), now-established quality and performance standards, and investment in manufacturing the product.

10.1 INTRODUCTION

Value-added wood products (VAWPs) are primary wood products that have been further processed into furniture, builders' joinery and carpentry (BJC), profiled wood, and engineered wood products (EWPs). EWPs include I-beams (also called I-joists) with their I-shaped cross-sections; finger-jointed sawnwood; glulam (sawnwood glued into beams); and laminated veneer lumber (LVL), which is formed by gluing together sheets of veneer and resawing to desired dimensions. Cross-laminated timber (CLT), although relatively new, is now a mainstream EWP.

10.2 IMPORTS OF VALUE-ADDED WOOD PRODUCTS

10.2.1 Wooden furniture

Global furniture production was $410 billion in 2015, down slightly from 2014. About 65% of furniture is consumed in the countries in which it is manufactured, and the other 35% is exported. The volume of furniture trade was estimated at $140 billion in 2015; this was a slight decline from 2014, but the overall trend is for increasing furniture exports (CSIL, 2016). Furniture manufacturing is labour-intensive, and it is increasingly taking place outside the UNECE region, even though manufacturing in the region has also increased in recent years.

Economic growth in the emerging markets has long outpaced that in the UNECE region. This has led to an increase in both labour costs and demand for furniture in emerging markets, thereby helping close the gap between those markets and the UNECE region. The global increase in furniture consumption is explained partly by increased local consumption in furniture-producing countries outside the UNECE region.

Furniture (both wooden and other materials) produced in the EU accounts for about 25% of global furniture production and 2-4% of the production value of the EU manufacturing sector, and it provides close to 1.1 million jobs (EFIC, 2016). Italy's furniture industry was composed of about 30,000 companies (180,000 employees) in 2015, and it generated revenue approaching $28 billion. In contrast, Germany's furniture industry had about 500 companies (83,700 employees) in 2015 and generated revenues of about $19.3 billion. Italy's furniture sector contracted by 30% between 2007 and 2014; it showed only modest growth (+3.6%) in 2015 and is still far below its peak. Germany returned to its 2008 production high in 2015 (Italy Europe 24, 2016).

The furniture industry has changed dramatically in recent decades and continues to do so. There are clear advantages in producing furniture in low-cost countries, but the cost difference is becoming less obvious as production technologies and cost structures change. Further automation, the better management of storage and working capital, the streamlining of logistics chains, and full customization to meet customer needs are some of the ways in which companies and industries can retain their competitiveness. Some leading furniture manufacturers are working to improve their ability to serve the marketplace in terms of quality, customization and quick delivery. Typical delivery times are now 10-12 weeks, but companies are working to cut that to 8-10 weeks. Some companies have launched "express lines" that are ready to deliver at any time. One technique for quick, custom-made delivery is to stock furniture in whitewood and finish after order, allowing the delivery of custom-made furniture in less than two weeks. Some manufacturers have specific "designer furniture collections" that follow the business models of other industries – such as the clothing industry, in which brands like H&M produce designer clothes for the masses.

The US is the world's largest furniture importer, with healthy growth in furniture demand and imports in the last few years. US furniture imports increased by 76% from 2009 to 2015 and now clearly exceed pre-global financial crisis (i.e. 2008) levels. Forecasts for 2016 are cautious but still point to growth; the latest forecast (made in April 2016) is for 3.9% growth in 2016, year-on-year (French, 2016).

As the baby-boomers retire and younger generations become the main consumer groups, buying patterns are changing. Furniture Today market research shows that online retail sales of furniture comprised 10% of the total US market in 2015 and 5% of the total revenue of traditional brick-and-mortar furniture stores. A significant proportion of online shoppers now use their mobile devices to buy furniture: in 2010, only 5% of consumers used a mobile device to access furniture stores online, but this had grown to 52% by 2015. On the other hand, over half of younger consumers still visited furniture stores in person before making purchases online. This trend has also been observed in other industries, with consumers still wanting to see products at bricks-and-mortar stores before making purchases online. A challenge for bricks-and-mortar furniture stores is to ensure that visitors ultimately make their purchases in their online stores, not those of competitors. Product customization is one way to take full advantage of online shopping, providing consumers with exactly what they want, keeping storage volumes down, and offering products at good prices.

Table 10.2.1 shows the value and market share of wooden furniture imports for the top five importing countries in 2014 and 2015; graph 10.1.1 shows the value of imports for those countries from 2011 to 2015. The US imported wooden furniture valued at $18.9 billion in 2015, 10% higher than in 2014. This was the fourth consecutive year of growth in US wooden furniture imports, showing the continued strength of that market. When measured in domestic currencies, the largest European furniture import markets grew at differing rates in 2015, with Germany and the UK reflecting stronger economies and France's market remaining steady over the year.

Currency depreciations, however, especially the weak euro, translated into much lower dollar values and accounted for all the contraction in the value of imports in euro economies; in the UK, the value of furniture imports was $4.3 billion (up by 1% in that currency). Japanese furniture imports contracted by more than 10% in US dollar terms. The French and German markets differ from other major import markets in that they are dominated by intraregional European imports, with 80% of their imported wooden furniture coming from other EU countries; in contrast, 75% of US imports and 50% of UK imports come from Asia. The leading Asian suppliers to the UNECE region are (in descending order, by value) China, Viet Nam and Indonesia.

GRAPH 10.1.1

Wooden furniture imports, top five importing countries, 2011-2015

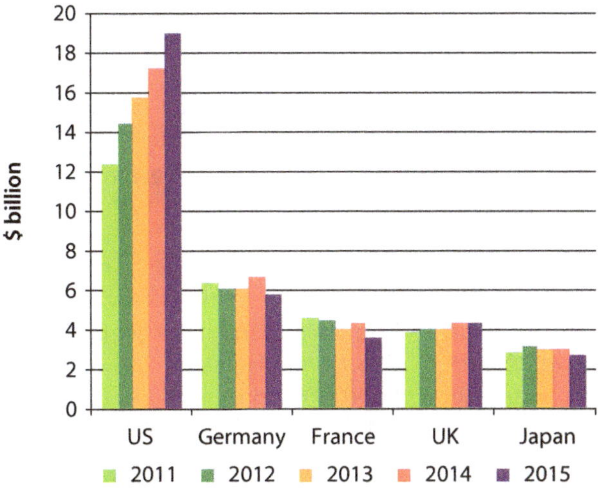

Sources: Eurostat, 2016; Trade Statistics of Japan, 2016; US International Trade Commission, 2016.

TABLE 10.2.1

Value of furniture imports and supplying regions by market share, top five importing countries, 2014 and 2015

(value in billion dollars, and market share in percentage)

	US		Germany		France		UK		Japan	
	2014	2015	2014	2015	2014	2015	2014	2015	2014	2015
Total value of imports	17.1	18.9	6.6	5.8	4.3	3.6	4.3	4.3	3.0	2.6
Of which furniture parts	2.4	2.7	1.3	1.2	0.8	0.7	0.8	0.8	0.6	0.5
Origin (%)										
Asia	73.1	74.3	16.3	16.4	21.1	22.1	51.2	51.2	88.2	88.7
Europe	11.1	10.2	83.2	83.1	77.7	76.3	45.6	45.4	11.0	10.4
North America	9.2	9.0	0.1	0.1	0.5	0.7	0.9	1.0	0.7	0.8
Latin America	6.4	6.4	0.2	0.2	0.4	0.5	2.1	2.2	0.1	0.1
Others	0.1	0.1	0.2	0.2	0.3	0.4	0.2	0.2	0.1	0.0

Sources: Eurostat, 2016; Trade Statistics of Japan, 2016; US International Trade Commission, 2016.

10.2.2 Trade policy issues in markets for value-added wood products

This Review has followed the antidumping case of Chinese-made bedroom furniture since 2005. The dispute began in 2004 when US manufacturers accused Chinese companies of unfair pricing. The US Department of Commerce imposed import duties on the products in 2005, ranging from 0% to more than 200%, effectively changing that part of the wood products trade between China and the US. The US International Trade Commission (ITC) has started its second five-year sunset review, and in October 2015 it announced it was reviewing whether duties on Chinese-made bedrooms were still warranted (Russell, 2016); some items were removed from the list in 2015. The outcome of the five-year review will be based largely on the ITC's analysis of whether removing the duties will lead to continued injury among US producers still manufacturing wooden bedroom items. Many in the industry argue that this segment is almost non-existent but, naturally, US-based companies will have their say on this.

The Trans-Pacific Partnership Agreement (TPP) is an initiative to remove tariffs on some 18,000 consumer products, including VAWPs, with the aim of increasing trade between parties (Russell, 2016). It was signed on 4 February 2016 in Auckland, New Zealand, after seven years of negotiation, but it has not entered into force. The main parties to the agreement (from the perspective of VAWPs) are Japan and the US as consumers and most Southeast Asian countries as producers.

The TPP is intended to deepen economic ties between parties by slashing tariffs and fostering trade to boost growth. Critics fear a continued loss of jobs, including in the furniture industry, and the ITC has organized hearings to alleviate these concerns in the US. Although China is not a party to the TPP, Malaysia and Viet Nam are. Both are large furniture exporters to the US, and furniture imports may again put pressure on slowly recovering US furniture manufacturers. The TPP aims to implement labour reforms and establish higher labour standards in all signatory countries, but the furniture industry does not believe this will have an impact (at least not immediately) on the difference in costs between producer and consumer countries.

10.2.3 Builders' joinery and carpentry, and profiled-wood markets

The markets for BJC have developed unevenly in the UNECE region. US imports have continued a strong increasing trend, reaching $2 billion in 2015. Imports by the main European consumer countries, Germany and France, were almost flat in euro terms (-0.7% and +0.4%, respectively) in 2015, but the strong euro devaluation against the dollar implied a contraction of 17% and 16%, respectively, in those two markets. UK imports grew slowly in US dollar terms, and the share of Asian-produced wood products in that market rose. Imports to Japan continued a dramatic slowing trend in 2015 (graph 10.2.2 and table 10.2.2).

The recently calibrated Leading Indicator of Remodeling Activity (LIRA) measures homeowner spending on home improvements in the US for the current quarter and provides forecasts for the subsequent three quarters. The LIRA showed a 4.3% gain in the first quarter of 2016, and the forecast in April 2016 was for 7.6% annual growth in 2016 (Harvard University, 2016). If this growth is achieved, remodelling spending would surpass (in nominal terms) the previous peak of 2006. Strongly accelerating growth in home improvement and repair spending is expected in 2017. LIRA forecasts that spending on home remodelling in the US will be up by 8.6%, year-on-year, in the last quarter of 2016 and by 9.7%, year-on-year, in the first quarter of 2017.

The remodelling market in the US has improved steadily in recent years, and a research analyst in the Remodeling Futures Program has said that homeowners are incorporating larger, more discretionary projects in their home-improvement priorities. Ongoing gains in home prices and sales are encouraging more homeowners to pursue larger-scale improvement projects in 2016, with permitted projects climbing at a good pace. This increased demand will soon require the industry to invest more in production capacity to ensure steady supply. The level of annual spending on repair and remodelling in the US is expected to reach nearly $325 billion in 2017. Up to 25% of all sawnwood and 16% of all structural and non-structural panels are consumed in remodelling.

Remodelling activity in the US did not decrease as much as new-home activity during the global financial crisis, but it was not immune to economic difficulty. Now, as customer confidence returns, remodelling activity is picking up at a faster pace than new-home construction. Also, some necessary maintenance of housing stock was not carried out during the crisis, and the need to catch up is helping grow the market for remodelling goods. Another reason for the growing market is the retirement of baby-boomers, who are remodelling houses to suit their needs in older age, especially for improved access.

The National Association of the Remodeling Industry publishes regular updates of remodelling market sentiment in its Remodeling Business Pulse Survey. The March 2016 survey pointed to a slowdown in the rate of remodelling growth and lower expectations of growth in the second quarter of 2016. Growth continues but with a slowing trend, contrary to the previously mentioned LIRA estimate. The dominant majority (70%) of market participants expect growth, but a small number (9%) sees some level of decline.

US imports of profiled woods have been growing, but the rate of growth is slowing and there has been recent softness in the market. The value of US imports was $1.2 billion in 2015 (up by 4% compared with 2014), still $500 million below the 2006 peak (in nominal terms), which may not be reached again for many years.

TABLE 10.2.2

Value of builders' joinery and carpentry imports, and supplying regions by market share, top five importing countries, 2014 and 2015

(values in billion dollars and market share in percentage)

	US		Germany		France		UK		Japan	
	2014	2015	2014	2015	2014	2015	2014	2015	2014	2015
Total value of imports	1.9	2.0	1.2	1.0	0.6	0.5	0.8	0.9	1.1	0.9
Origin (%)										
Asia	33.1	32.5	7.5	7.5	9.4	8.5	36.6	37.7	91.6	93.1
North America	47.3	48.7	0.2	0.2	0.5	0.8	1.9	2.0	2.9	2.6
Europe	4.8	4.7	91.5	91.2	87.8	88.5	57.4	56.3	3.6	2.7
Latin America	14.1	13.7	0.3	0.3	1.2	1.0	2.9	2.6	0.0	0.0
Others	0.6	0.5	0.6	0.8	1.1	1.2	1.3	1.4	1.9	1.5

Sources: Eurostat, 2016; Trade Statistics of Japan, 2016; US International Trade Commission, 2016.

GRAPH 10.2.2

Builders' joinery and carpentry imports, top five importing countries, 2011-2015

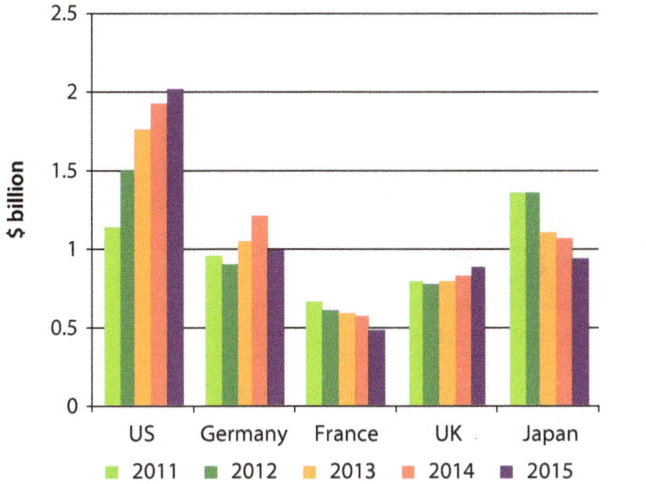

Sources: Eurostat, 2015; Trade Statistics of Japan, 2015; US International Trade Commission, 2015.

Source: Harry Bagley, 2016.

Many larger suppliers to the US markets have made permanent changes to production capacity, and a large share of the existing production is now targeted at other markets, principally China. Adding production capacity is relatively easy, but rebuilding sales organization and distribution channels, especially for retail customers, takes more time and is unlikely to happen at the same scale as before, even if the market grows faster again. Those suppliers continue to have access to sustainable raw materials, however, and they are benefiting from the strong US dollar. Major players in the US softwood moulding markets are Brazil (34% market share), Chile (28%), Canada (13%), China (11%), Mexico (9%) and Argentina (3%). These six countries comprise 97% of the total import market for softwood mouldings. Profiled-wood markets in Europe declined in 2015 and are supplied mainly by regional producers (graph 10.2.3 and table 10.2.3).

GRAPH 10.2.3

Profiled-wood imports, top five importing countries, 2011-2015

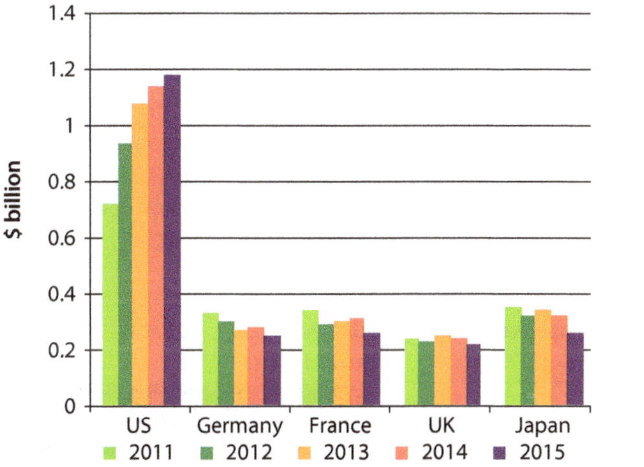

Sources: Eurostat, 2016; Trade Statistics of Japan, 2016; US International Trade Commission, 2016.

TABLE 10.2.3

Profiled-wood imports and supplying regions by market share, top five importing countries, 2014 and 2015

(values in billion dollars and market shares in percentage)

	US		*Germany*		*France*		*UK*		*Japan*	
	2014	*2015*	*2014*	*2015*	*2014*	*2015*	*2014*	*2015*	*2014*	*2015*
Total value of imports	1.1	1.2	0.3	0.2	0.3	0.3	0.2	0.2	0.3	0.3
Origin (%)										
Asia	19.6	22.4	21.2	22.9	9.2	8.8	55.2	62.7	76.4	78.3
North America	10.9	11.6	1.1	0.9	0.6	0.6	4.9	4.0	7.1	7.6
Europe	3.1	3.6	72.1	69.3	69.9	63.2	38.2	31.4	10.2	8.0
Latin America	65.0	61.8	3.1	4.6	19.0	26.3	1.5	1.7	5.0	4.2
Others	1.3	0.6	2.5	2.3	1.3	1.1	0.2	0.2	1.3	1.8

Sources: Eurostat, 2015; Trade Statistics of Japan, 2015; US International Trade Commission, 2016.

10.3 ENGINEERED WOOD PRODUCTS

Most EWPs covered in this chapter are heavily dependent on residential construction (new and, just as importantly, repair and renovations) and increasingly on non-residential building construction, including schools, restaurants, stores and warehouses.

10.3.1 Glulam timber

10.3.1.1 Europe

Comprehensive data on the production and consumption of glulam in Europe are currently unavailable, but some information exists on trade and production at the country level.

Austria was the largest producer of glulam in Europe in 2014 (the most recent year for which production data could be found), at roughly 1.5 million m^3, but production has been contracting since 2012, due primarily to reduced construction in southern Europe (a key export destination for Austrian glulam). Several Austrian glulam producers have shut down or reduced production as a result (Timber-online.net, 2015). Japan – another major importer of Austrian glulam – reduced its overall imports of glulam in 2015 by 3% and the Austrian component by 22% (table 10.3.1) (Timber-online.net, 2016b).

The majority of Japan's imports of glulam come from European sources; while Austria lost market share, Finland strengthened its status as the leading exporter of glulam to Japan, and Romania's glulam exports to Japan increased by 29%. In April 2016, the average import price (CIF) for European glulam delivered to Japan was about \$474/m^3 (Timber-online.net, 2016c).

TABLE 10.3.1

Japan's imports of laminated timber, 2014-2015

(thousand m^3)

	2014	2015	Change (%) 2014-2015
Finland	233	242	3.9
Romania	131	168	28.2
Austria	136	107	-21.3
Estonia	57	58	1.8
Russian Federation	43	52	20.9
Sweden	51	44	-13.7
China	56	17	-69.6
Germany	16	14	-12.5
Total	723	702	-2.9

Note: "Laminated timber" includes both glulam and cross laminated timber.
Source: Timber-online.net, 2016b.

Germany's glulam industry is oriented towards its large domestic market, and indications are that the industry is holding steady as a result of increasing domestic construction of larger residential buildings. The share of wooden buildings is also increasing, due partly to increased demand for shelters for asylum seekers.[24] Several companies are already investing – or

Source: Harry Bagley, 2016.

are planning to invest – in improving their facilities (Timber-online.net, 2015).

The Swedish Forest Industry Federation reported that glulam production there set a record in 2015, and further growth is expected in 2016. Trends in green building and increased domestic demand is credited for this development (Timber-online.net, 2016a).

EU imports of glulam from outside the subregion are increasing, although the overall share of total EU consumption is still small (about 5%) (table 10.3.2). Much of the imported glulam consists of LVL for window manufacturing. The most important sources of EU28 imports from outside the subregion are, in descending order by volume, Indonesia, Malaysia, China and the Russian Federation; most glulam imports from China and the Russian Federation comprise pine and larch, and imports from Indonesia and Malaysia consist of meranti (Global Wood, 2015).

TABLE 10.3.2

EU28 imports of glulam, 2013-2015

(\$ million)

	2013	2014	2015	Change (%) 2014-2015
Europe	739.8	723.2	623.0	-13.9
Asia	96.5	100.1	105.4	5.3
Indonesia	30.6	34.3	41.2	20.2
Malaysia	26.6	31.8	31.2	-1.9
China	23.2	21.9	23.3	6.6
Russian Federation	16.9	22.4	22.1	-1.0
Latin America	5.0	4.7	5.0	6.2
North America	3.6	5.5	4.5	-17.7
Others	6.3	4.6	4.8	3.9

Note: Exchange rate from euros to US dollars: 2013 = 1.3281, 2014 = 1.3285, 2015 = 1.1095.
Source: Eurostat, 2016.

[24] According to a survey conducted by Heinze in Germany, wood shelters account for 38% of the new housing provided to immigrants (Timber-online.net, 2016d).

10.3.1.2 North America

Glulam consumption in North America is highly dependent on new-building construction and renovation, with only about 4% used for industrial or other applications (graph 10.3.1). Overall production of North American glulam timber declined from 750,000 m^3 in 2006 to 285,000 m^3 in 2009. Production has increased steadily since then, however, with a forecast production of 446,000 m^3 in 2016 (graph 10.3.2 and table 10.3.3).

GRAPH 10.3.1

Uses of glulam, North America, 2015

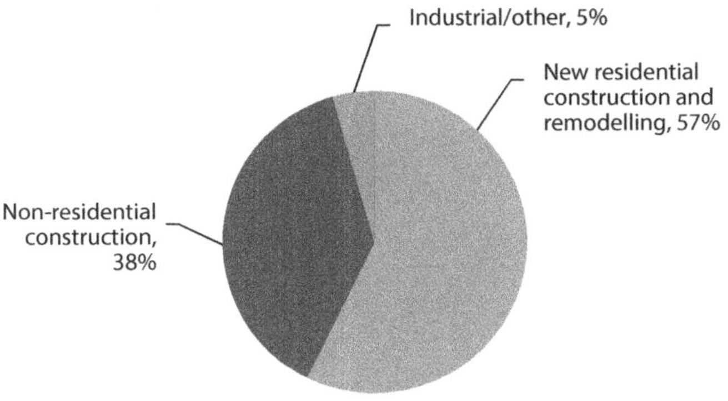

Source: APA, 2016.

GRAPH 10.3.2

Glulam production, North America, 2008-2016

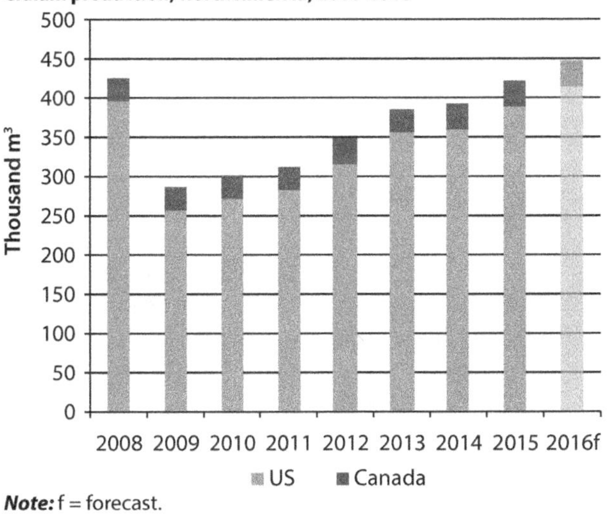

Note: f = forecast.
Source: APA, 2016.

TABLE 10.3.3

Glulam production and consumption, North America, 2014-2016

(thousand m^3)

	2014	2015	2016f	Change (%) 2014-2015
US				
Production	358.5	387.7	412.3	8.1
Consumption				
Residential	190.8	216.9	236.9	13.7
Non-residential	136.9	144.6	150.8	5.6
Industrial, other	16.9	16.9	18.5	0.0
Total consumption	344.6	378.5	406.2	9.8
Inventory change	13.8	9.2	6.2	-33.3
Canada				
Production	33.8	32.3	33.8	-4.4
North America				
Total production	392.3	420.0	446.2	7.1

Notes: f = forecast. Conversion factor: 1 m^3 = 650 board feet. Canadian imports are assumed to be minimal.
Source: APA, 2016.

10.3.2 I-beams

I-beams are more than 80% dependent on new-home construction, mostly single-family homes. Builder surveys indicate that the I-beam share of raised wood-floor area (not including concrete floor area) varied between 48% and 54% in the five years to 2013 (graph 10.3.3). The I-beam market share was only 16% in 1992.

GRAPH 10.3.3

I-beam market share of total raised-wood floor area, single-family homes, US, 2007-2013

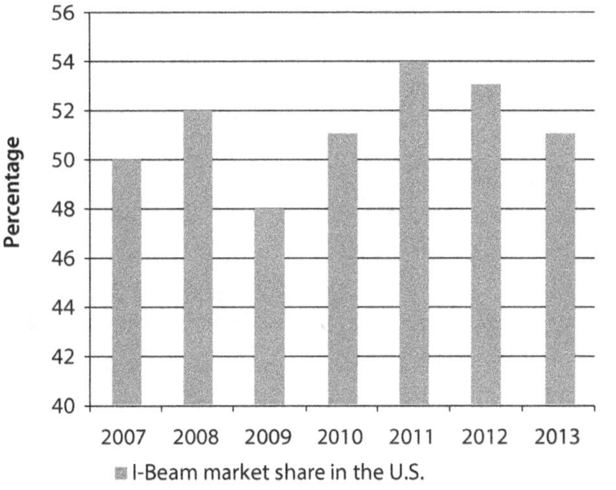

Source: National Association of Home Builders' Home Innovation Research Lab surveys.

Demand for I-beams peaked in 2005, which equated to the practical capacity of I-beam plants at that time; housing starts were so high that manufacturers were producing all they could.

I-beam demand and production declined in 2009 when the US housing bubble burst. Roughly 115 million linear metres were produced in that year, and there have been significant increases since; the forecast for 2016 is 221 million linear metres, which would be a 91% increase over 2009 (graph 10.3.4 and table 10.3.4).

GRAPH 10.3.4

I-beam production, North America, 2008-2016

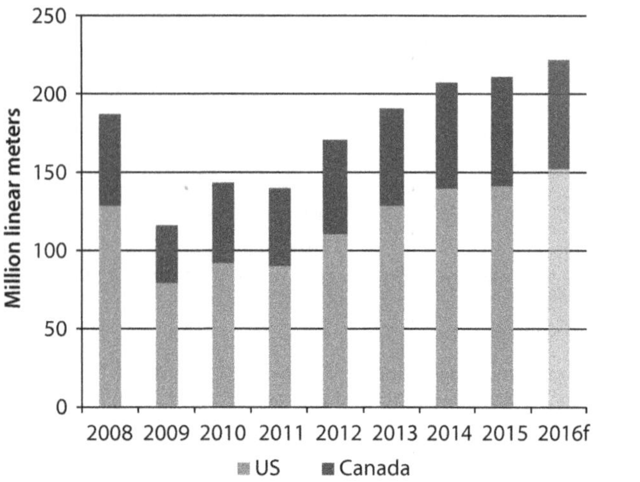

Note: f = forecast. Conversion factor: 1 linear metre = 3.28 linear feet.
Source: APA, 2016.

TABLE 10.3.4

Wooden I-beam consumption and production, North America, 2013-2015

(thousand m^3)

	2014	*2015*	*2016f*	*Change (%) 2014-2015*
US				
Production	139.3	141.8	151.5	1.8
Consumption				
New-residential	144.5	151.2	164.0	4.6
Repair, remodelling	4.0	4.3	4.3	7.5
Non-residential, other	10.1	11.0	11.6	8.9
Total consumption	158.5	166.5	179.9	5.0
Canada				
Production	67.1	68.0	69.5	1.3
Consumption	30.2	31.7	31.4	5.0
Inventory change	6.1	0.6	0.0	-90.2
North America				
Total production	206.4	209.8	221.0	1.6

Notes: f = forecast. Conversion: 1 linear metre = 3.28 linear feet.
Source: APA, 2016.

Source: APA, 2016.

More than 90% of I-beams are used in new residential construction (graph 10.3.5), with the balance used in non-residential building construction and in repair and remodelling.

GRAPH 10.3.5

I-beam end-uses in North America, 2015

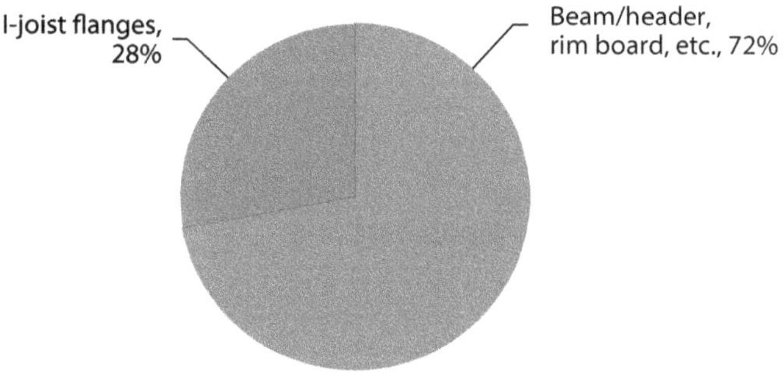

Source: APA, 2016.

10.3.3 Laminated veneer lumber

Most LVL is used in new-home construction. In 2015, 72% of total consumption was in beams and headers, rim boards and like applications, and the balance was in I-joist flanges (graph 10.3.6). Rim boards are used on the perimeters of I-beam floor systems to provide fastening points for the I-beams and to assist in distributing wall loads.

GRAPH 10.3.6

Laminated veneer lumber end-uses in North America

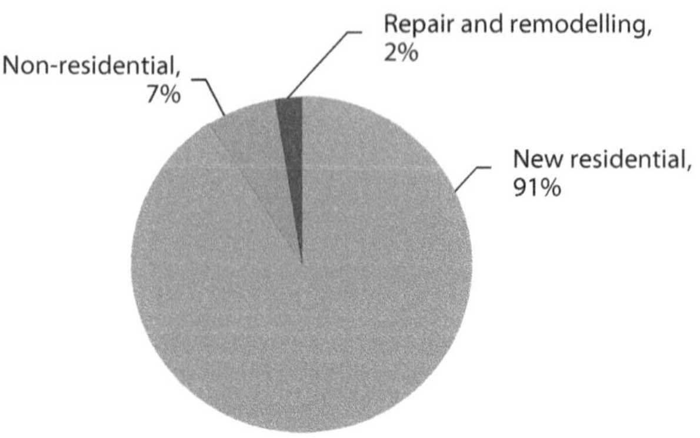

Source: APA, 2016.

North American production of LVL peaked with the US housing market in 2005 at 2.6 million m³ and declined thereafter to 2009, along with I-beam production. According to forecasts, 2.04 million m³ of LVL will be produced in North America in 2016, up by 121% from the 2009 trough (graph 10.3.7 and table 10.3.5).

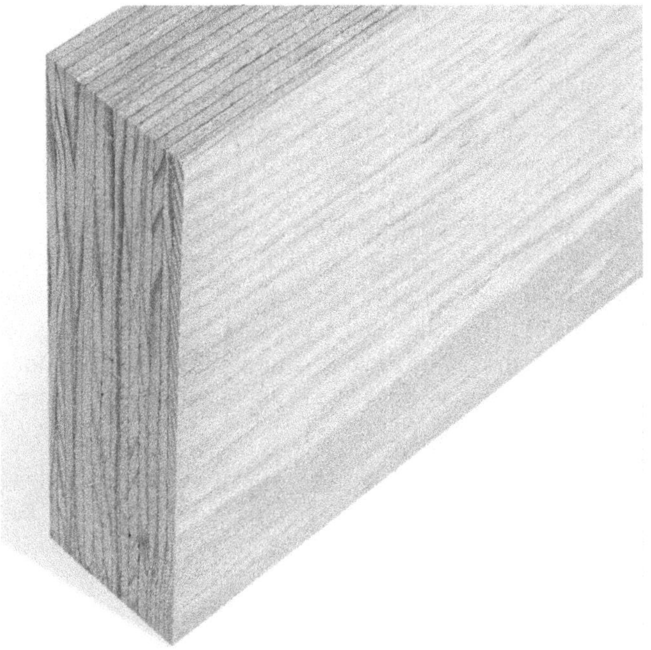

Source: APA, 2016.

GRAPH 10.3.7

Laminated veneer lumber production in North America, 2008-2016

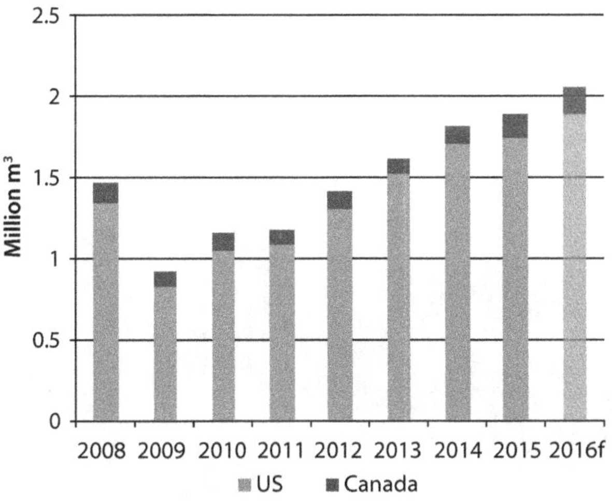

Notes: f = forecast. Conversion factor: 1 m³ = 35.315 cubic feet.
Source: APA, 2016.

LVL is well-accepted for use in beams and headers, and consumption should grow as the housing market improves. Like other EWPs, LVL allows the use of longer spans and fewer pieces to carry the same loads as conventional wood products.

In addition to the EWPs discussed above, a number of other structural composite lumber products are manufactured in North America, including parallel strand lumber (PSL), laminated strand lumber (LSL) and oriented strand lumber (OSL). These products are made using strands of wood of varying lengths and widths to achieve differing strength and stiffness properties. PSL and LSL have been manufactured for several years, primarily by one company, and production volumes are low compared with other EWPs. Uses for OSL are expected to be the same as for solid sawn lumber and glulam – posts, beams, headers, rim boards and structural framing lumber.

TABLE 10.3.5

Laminated veneer lumber consumption and production in North America, 2014-2016

(thousand m³)

	2014	*2015*	*2016f*	*Change (%) 2014-2015*
Demand				
I-beam flanges	515	527	558	2.3
Beams, headers, others	1294	1356	1487	4.8
Total consumption	1810	1883	2045	4.0
Production				
US	1699	1739	1883	2.4
Canada	110	144	161	30.9
Total production	1810	1883	2045	4.0

Notes: f = forecast. Conversion: 1 m³ = 35.3137 cubic feet.
Source: APA, 2016.

10.3.4 Cross-laminated timber

10.3.4.1 Europe

CLT production is expanding globally, with new production facilities constructed outside the DACH countries (Germany, Austria and Switzerland), where production first started. Global production in 2015 was estimated at 650,000-700,000 m³ (Plackner, 2015), and the volume is set to increase to 1 million m³ in 2016 as investments in Finland, Japan, Latvia and the US come online.

Source: proHolz, 2014.

CLT production is still concentrated in Europe. Growth in DACH countries has slowed considerably since 2011, however, with no further big investments planned there, and most of the growth coming from the modernization and expansion of existing facilities (graph 10.3.8).

GRAPH 10.3.8

Cross-laminated timber production in the DACH countries, 2008-2015

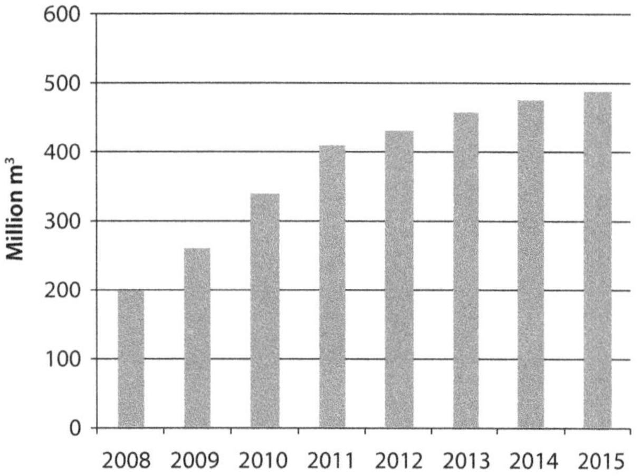

Note: DACH countries are Germany, Austria and Switzerland.
Source: Holzkurier, 2015.

With the continuing weakness of European construction markets, the full development of international standardization is widely acknowledged as a potential way of reducing production costs for consumers and increasing demand. CLT construction is currently about 10-15% more expensive than masonry and cement construction (Timber-Online, 2016e). A reduction in CLT costs would enable the sector to take full advantage of the benefits of the material, such as the speed and efficiency of installation, design versatility, reduced waste, lighter weight (compared with concrete), and energy efficiency.

10.3.4.2 North America and Japan

CLT-importing countries such as Japan and the US are investing heavily in their own production capacity. CLT is gaining interest in North America in both the construction and wood industries, and it is expected that production and consumption will increase in coming years. In 2011, this Review discussed the premiere of CLT in Canada. Today, three commercial CLT producers in Canada are making a significant amount of product for non-residential construction and industrial applications. In the US, the Smartlam CLT plant has announced plans to quadruple production in 2016, to an annual capacity of 116,000 m^3, which will make it the biggest CLT production site globally. The D.R. Johnson Lumber Co. in Riddle, Oregon, has invested in the production of CLT and is enthusiastic about its future in North America. North American CLT production has mostly been used as temporary roading in the fracking and oil industries and for forestry work and it is not yet common in residential construction. Nevertheless, the use of CLT for building is quickly gaining momentum, thanks in large part to interest shown in west-coast cities with strong wood cultures (e.g. Portland, Seattle and Vancouver) and the rapid quality-and-performance standardization process.

In 2012 the American National Standards Institute approved the ANSI/APA PRG 320-2012 Standard for Performance-Rated Cross-Laminated Timber, a product standard that details manufacturing and performance requirements for qualification and quality assurance. CLT was included in the International Building Code in 2015 (Showalter, 2015).

In Japan, interest in CLT is driven partly by the product's good seismic performance. There were three CLT production facilities in Japan in 2015, with a total annual output of 10,000 m^3. Annual output is projected to increase to 500,000 m^3 in the next ten years.

CLT has the potential to change the way in which wood is used in construction. The manufacturing process for CLT lends itself to prefabrication; for example, the product dimensions can easily be custom-made for particular applications, and windows and other openings can be pre-made at the factory using computer numerical control technology. The use of CLT often shortens construction times, and CLT provides excellent environmental performance; importantly, CLT has the ability to compete with concrete and steel for larger, taller, multi-story construction applications. All these attributes have contributed to its growth.

In spring 2016, construction began on an 18-storey wood hybrid student residence high-rise at the University of British Columbia in Vancouver, Canada. At 53 metres in height, it will be one of the tallest wooden buildings globally, consisting of a mass timber superstructure on top of a concrete base and central concrete towers. There are already many tall, multi-story, CLT-based structures in existence, and many other architecturally ambitious projects are in the planning stage.

Source: Acton Ostry Architects, 2016.

10.4 REFERENCES

APA (The Engineered Wood Association). 2016. Structural panel and engineered wood yearbook. APA Economics Report E175. Available at: www.apawood.org

CSIL. 2016. World furniture outlook 2016. Centre for Industrial Studies (CSIL). Available at: www.csilmilano.com

EFIC. 2016. European Furniture Industries Confederation (EFIC). Available at: http://www.efic.eu/default.aspx

Eurostat. 2016. External trade. Available at: http://epp.eurostat.ec.europa.eu

French, D. 2016. Furniture Today. Available at: www.furnituretoday.com

Global Wood. 2015. Wood products prices in Europe, 1-15 October 2015. Available at: www.globalwood.org/market/timber_prices_2015/aaw20151001e.htm

Harvard University. 2016. Remodeling Futures Program at the Joint Center for Housing Studies, Harvard University. Available at: www.jchs.harvard.edu/research/remodeling-futures

Italy Europe 24. 2016 Despite Italian dominance, German furniture sector is a model in many respects. 16 April 2016. Available at: www.italy24.ilsole24ore.com

Plackner, H. 2015. Brettsperrholz wächst global 2015. Holzkurier heft, 12-13. Available at: www.timber-online.net

Russell, T. 2016. Furniture Today. Available at: www.furnituretoday.com

Showalter, B. 2015. What the 2015 international building code means for wood construction: Part I. In: The construction specifier 2015. Available at: www.constructionspecifier.com/what-the-2015-international-building-code-means-for-wood-construction-part-i

Timber-online.net. 2015. Further decrease in glulam sector. 7 September 2015. Available at: www.timber-online.net

Timber-online.net. 2016a. Record figures for Swedish glulam sales in 2015. 23 February 2016. Available at: www.timber-online.net

Timber-online.net. 2016b. Japan imported less laminated timber in 2015. 3 March 2016. Available at: www.timber-online.net

Timber-online.net. 2016c. Japanese laminated timber imports increased. 13 June 2016. Available at: www.timber-online.net

Timber-online.net. 2016d. Deutsche Holzbauquote auf 16% gestiegen. 24 May 2016. Available at: www.timber-online.net

Timber-online.net. 2016e. BSP muss international standardisiert werden 21.04.2016. Available at: www.timber-online.net

11 HOUSING AND CONSTRUCTION

Lead author: ***Delton Alderman***

HIGHLIGHTS

- The European and North American housing markets are still recovering from the 2008 global financial crisis.
- In aggregate, advanced economies have been sluggish, and unemployment remains high in several countries. As a result, the housing and construction markets have been muted in the UNECE region.
- Construction increased by 0.1% in the EU28 between March 2015 and March 2016. The gain was due primarily to improvements in the building construction sector, but the civil engineering sector declined, year-on-year.
- On a monetary basis, remodelling is the largest component of euro-area residential construction. New residential construction, however, is forecast to grow at a faster rate than remodelling in the immediate future.
- Housing completions in the Russian Federation decreased by 1.1% in 2015, with just over 280,000 new residential dwellings put in place.
- All sectors of the US housing market improved in 2015. Beginner or starter housing has been weak, however, and the number of dwellings built is insufficient to match population growth.
- The number of negative-equity houses is at long-term historical levels and is not expected to hinder the US housing market in the near future, as it has done in the last few years.
- US household formations have improved but are still less than historical averages.
- Investor purchases of US housing remain a strong influence in the existing-housing sales sector but are not the factor they were at the beginning of the housing recovery.
- Canada's economy stalled in 2015 but is expected to improve in 2017 and 2018; housing starts are expected to decline slightly in 2016 and 2017.

11.1 INTRODUCTION

The economic state of countries has a significant effect on housing and construction markets. Historically, housing construction and sales have increased in good economic times as consumers have built or purchased houses. The opposite is also true: in less prosperous economic cycles, housing construction and sales typically decline.

The International Monetary Fund (IMF) and the World Bank foresee subdued economies in Europe and the US in 2016 and 2017 (World Bank, 2016). For the euro area, GDP is forecast to grow at 1.5% in 2016 and 1.6% in 2017, and the forecast for global GDP growth is 3.2% in 2016 and 3.5% in 2017. Forecasts are higher for the US, at 2.4% in 2016 and 2.5% in 2017. In the Russian Federation, GDP growth was downgraded in 2015 to 3.7%, and the IMF has predicted it will be -1.8% in 2016 and 0.8% in 2017 (IMF, 2016). These forecasts suggest that housing construction and sales in 2016 and 2017 might be similar to those reported in the past few years.

11.2 EUROPE

11.2.1 Review and outlook

In 2014, 70% of EU28 citizens lived in owner-occupied houses, and the remainder were renters. Slightly more than 27% of citizens lived in owner-occupied homes with outstanding loans or mortgages, and about 43% lived in owner-occupied homes without loans or mortgages (Eurostat, 2016). Nearly 40% of EU28 citizens lived in flats; 25.6% were in semi-detached houses; and 33.7% were in detached houses.

The Euroconstruct [25] region's housing market is still influenced by the aftermath of the global financial crisis and other factors, and the economic forecast is for "subdued" growth in 2016 and beyond. The residential sector remains a vital component of the economy in the euro area, generating more than 46% of overall growth in the construction market (Euroconstruct, 2015). Allen (2016) reported that construction increased by 0.1% in both the euro area and the EU28 between March 2015 and March 2016. The gains in both groupings were due primarily to improvements in the building construction sector, but the civil engineering sector declined, year-on-year. On a monetary basis, remodelling is the largest component (about 60%) of euro-area residential construction. New residential construction is forecast to increase more than remodelling (in percentage terms) in the immediate future, however. In 2015, new residential remodelling accounted for 35.5% of total new construction spending, new non-residential construction for 21.6%, new residential housing for 24.4%, and civil engineering for 18.5% (Euroconstruct, 2015).

Source: proHolz, 2016.

The key factors positively affecting the residential sector are financing and improved economic prospects, consistent with a slowly recovering European economy. Through 2018, new residential spending is forecast to increase by 2.4% annually, with civil engineering increasing by 3.2% and non-residential by 2.3% per year. Residential spending is projected to decrease in Europe in the longer term, however, beginning in 2018, as markets mature in the larger European countries. Not only has demand composition changed recently (e.g., flats versus 1+2 family dwellings[26]), so too has supply. One facet of the new residential market is production; that is, the composition of the new-home subsector. New-home production is forecast to increase rapidly but will vary by country. Belgium, France, Germany, the Netherlands and Spain currently account for 73% of total residential output, and these countries are forecast to lead housing starts in the near term. The renovation of buildings to reflect changing demand requirements and preferences in light of demographic ageing will also become an essential aspect of future housing construction (Euroconstruct, 2015).

11.2.2 New housing

It is projected that there will be 1.48 million new housing permits and 1.19 million new-home starts in the Euroconstruct region in 2016 – roughly the same as in 2011 (Euroconstruct, 2015). A record 2.78 million homes were permitted in 2006,

[25] The Euroconstruct region comprises 19 countries. The western subregion consists of Austria, Belgium, Denmark, Finland, France, Germany, Ireland, Italy, the Netherlands, Norway, Portugal, Spain, Sweden, Switzerland and the UK. The eastern subregion comprises the Czech Republic, Hungary, Poland and Slovakia.

[26] 1+2 family dwellings are detached or semi-detached single or multi-unit structures for families.

compared with about 1.40 million units in 2015, a 49.7% decrease. It is forecast that nearly 655,600 flats and 528,500 1+2 family dwellings will be started, in 2016 and 812,100 flats and 661,600 1+2 family dwellings will be completed (graph 11.2.1) (Euroconstruct, 2015).

GRAPH 11.2.1

Building permits, housing starts and completions, Euroconstruct region, 2007-2018

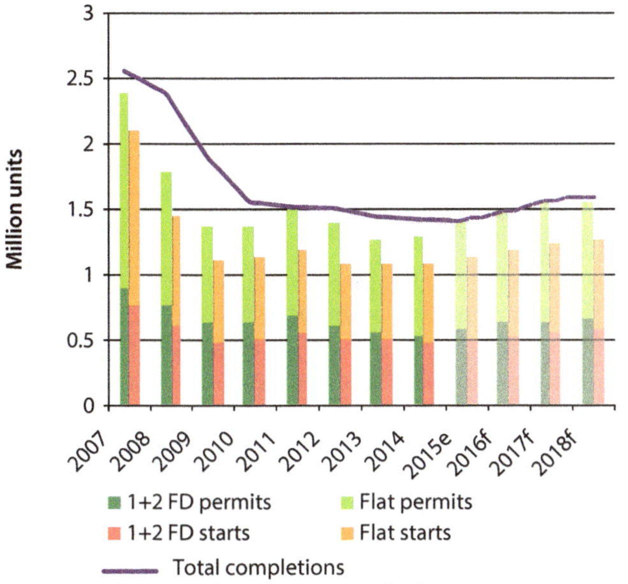

Notes: FD = family dwellings; e =estimate; f = forecast.
Source: Euroconstruct, 2015.

Table 11.2.1 presents the top five countries for new construction and remodelling, by value, in 2015 and 2016, with projections for 2017 and 2018. Germany ranks first in both new construction and remodelling in each of the four years (Euroconstruct, 2015).

TABLE 11.2.1

Top five Euroconstruct region countries for new construction and renovation expenditure, 2015-2018

(€ billion)

	2015e	*2016f*	*2017f*	*2018f*
New construction				
Germany	49.6	53.3	55.7	56.5
France	43.7	48.8	51.9	54.6
UK	42.2	43.2	44.5	44.1
Switzerland	19.2	19.4	18.8	18.1
Spain	18.8	20.7	21.8	22.6
Residential renovation				
Germany	114.5	114.5	114.5	114.5
Italy	64.9	65.7	66.1	66.7
France	53.6	54.7	55.5	55.3
UK	38.7	39.8	40.4	39.6
Spain	14.1	14.5	15.1	15.5

Notes: e = estimate; f = forecast.
Source: Euroconstruct, 2015.

11.2.3 Non-residential buildings and civil engineering

Demand for non-residential construction in the Euroconstruct region is affected by overall economic conditions and government funding (e.g., for buildings for education and health). Non-residential construction is forecast to expand by 2.2% in the Euroconstruct region in 2017 and by 1.9% in 2018. More than one-fifth of new non-residential construction consists of buildings for education (13%) and health (8%). Industrial and storage buildings are affected primarily by the macroeconomic environment. Specifically, industrial building is correlated with the condition of manufacturing and capacity utilization; domestic demand; and the export market. Storage relates to distribution requirements and logistics facilities, which are affected by changing consumer-retailing models. Economic conditions have a direct impact on commercial and office construction (Euroconstruct, 2015).

The aggregate cumulative forecast for growth in new non-residential construction in the Euroconstruct region for 2016-2018 is, by subsector: commercial – 12.3%; office – 10.9%; industry – 9.9%; health – 9.6%; storage – 8.8%; agriculture – 7.5%; and education – 5.5% (Euroconstruct, 2015).

New non-residential construction comprised 21.6% of new construction spending in the Euroconstruct region in 2015. New non-residential construction is predicted to increase by 3.8% in 2016, by 3.0% in 2017, and by 2.1% in 2018 (table 11.2.2). The UK, Germany, France, Spain and Italy (in descending order) were the five largest non-residential markets in 2013 (Euroconstruct, 2015).

TABLE 11.2.2

Non-residential construction spending forecast, Euroconstruct region, 2016-2018

(€ billion)

				Change (%)		
	2016f	*2017f*	*2018f*	2015-2016	*2016-2017*	*2017-2018*
New construction	233.8	240.8	245.8	3.8	3.0	2.1
Renovation	214.4	217.2	220.9	1.9	1.3	1.7
Total	448.2	458	466.7	2.9	2.2	1.9

Notes: e = estimate; f = forecast.
Source: Euroconstruct, 2015.

Civil engineering projects are affected by different factors in different countries: they are financed primarily by the public sector and "... the main factors of influence for the civil engineering demand are more related to politics and less to economic factors compared to the situation in other construction sectors" (Euroconstruct, 2015). Civil engineering started to improve in 2014 and continued to improve in 2015 (+3.3% over 2014). Projections for 2016 indicate that civil engineering spending will continue at about the same growth rate, to about €318.6 billion in 2016 (table 11.2.3; graph 11.2.2). Modest spending increases are forecast through 2018

(Euroconstruct, 2015). Germany, France, the UK, Italy and Spain (in descending order) were the five largest civil engineering markets in 2014.

TABLE 11.2.3

Civil engineering construction spending forecast, Euroconstruct region, 2016-2018

(€ billion)

	New civil engineering construction	*Civil engineering renovation*	*Total civil engineering*
2016f	198.7	119.9	318.6
2017f	209.3	122.7	332.0
2018f	216.3	125.1	341.4

Note: In 2014 prices. f = forecast.
Source: Euroconstruct, 2015.

GRAPH 11.2.2

European construction spending, 2007-2018

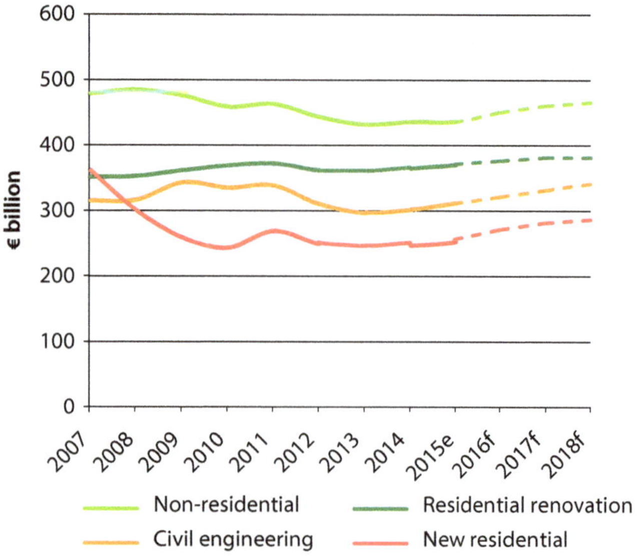

Notes: e = estimate; f = forecast.
Sources: Euroconstruct, 2009, 2013, 2015.

11.2.4 Residential construction and renovation

The value of new residential construction is projected to increase by 4.1% in 2017 and by 6.8% in 2018, year-on-year. The forecast value of total residential construction (new residential plus residential renovation) in 2016 is €644.7 billion, increasing to €670.1 billion in 2018 (table 11.2.4). On average, total new residential construction is forecast to increase by 3.9% (in nominal terms) from 2016 to 2018 (Euroconstruct, 2015).

Residential renovation is forecast to remain the principal construction activity in the euro area, increasing from €375.1 billion in 2016 to €382.2 billion in 2018. Housing renovation is forecast to increase by 1.1% in 2017 and by 1.9% in 2018. Home renovation projects have historically been supported by government programmes (Euroconstruct, 2015).

TABLE 11.2.4

Residential new construction and renovation spending forecast, Euroconstruct region, 2016-2018

(€ billion)

	New construction	*Renovation*	*Total residential*
2016f	269.5	375.1	644.7
2017f	280.5	379.3	659.8
2018f	287.8	382.2	670.1

Note: In 2014 prices. f = forecast.
Source: Euroconstruct, 2015.

11.2.5 Construction sector shares and growth: Contrasting western and eastern Europe

In the Euroconstruct's western subregion, total residential construction is predicted to increase from €257.8 billion in 2016 to €274.8 billion in 2018. Residential construction spending in the Euroconstruct's eastern subregion is forecast to increase from €11.7 billion to €13.0 billion in the same period (Euroconstruct, 2015).

New residential construction is the leading sector in terms of spending in the Euroconstruct's western subregion (39.0% of total construction spending), followed by new non-residential building (32.7%) and civil engineering (28.3%). Spending in the eastern subregion amounted to 7.4% of total European construction expenditure. New civil engineering and new non-residential construction accounted for 77.4% of new construction expenditure in the subregion and new residential construction for the remainder (22.6%) (graph 11.2.3) (Euroconstruct, 2015).

GRAPH 11.2.3

Share of new construction, by Euroconstruct subregion and sector, 2015

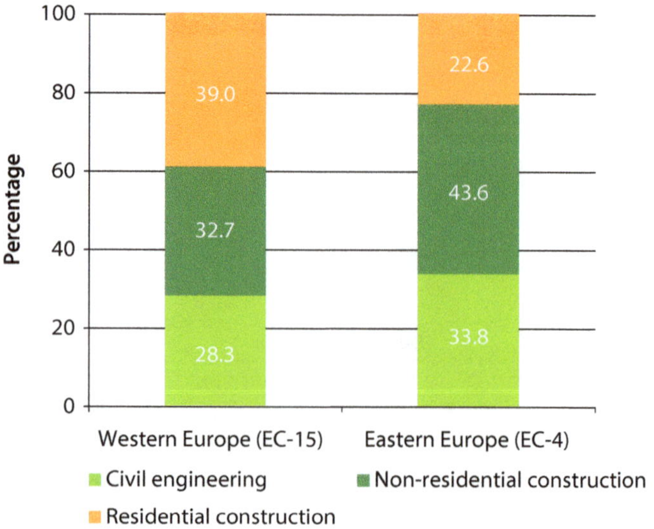

Source: Euroconstruct, 2015.

11.3 COMMONWEALTH OF INDEPENDENT STATES, WITH A FOCUS ON THE RUSSIAN FEDERATION

11.3.1 Housing construction in the Russian Federation, 2015-2016

Just over 280,000 residential buildings were built in the Russian Federation in 2015, a decrease of 1.1% over 2014. Overall, 418.2 million m² of floor space was put in place, an increase of 3.4%, year-on-year, indicating an increase in the average floor area of new buildings (Federal State Statistics Service, 2016a). Private developers built 264,000 units of the total, with an area of 34.3 million m². This was nearly 0.4% less than the area commissioned in 2014 (Federal State Statistics Service, 2016b).

Approximately 1.17 million apartments were commissioned in the Russian Federation in 2015, with 83.8 million m² of residential space constructed, a decrease of 0.5% compared with 2014. According to PMR (2016), subdued near-term economic growth is expected to result in sizeable budget cuts and delays in residential construction programmes.

11.4 NORTH AMERICA

The US housing market has continued its ascent from the 2009 trough, while the Canadian market has remained steady despite recent declines in oil prices and revenues (recently a strong contributor to the Canadian economy) (graph 11.4.1). The housing markets in the US and Canada are both still recovering from the housing crash and the global financial crisis. Both have improved but there is room for much more, particularly in the US.

GRAPH 11.4.1

Housing starts, North America, 2000-2018

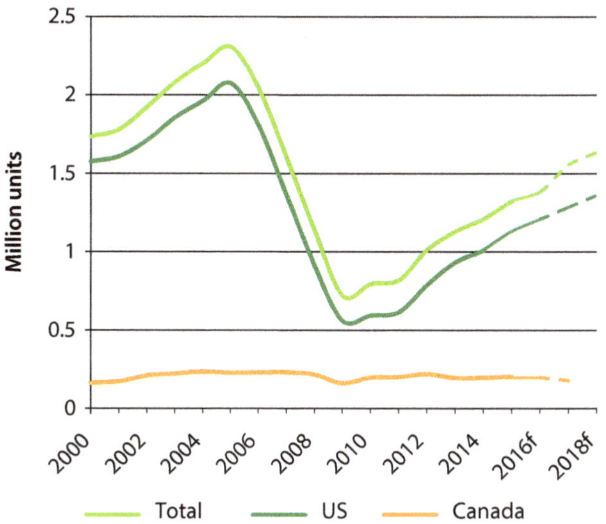

Notes: e = estimate; f = forecast. 2018 forecast for Canada is missing.
Sources: US Census Bureau, 2016a; CMHC, 2016; Mortgage Bankers Association, 2016.

11.4.1 US housing market

US housing starts increased by 11% in 2015, but yet annual starts were lower than the 1963-to-2008 average of 1.46 million units. On a per capita basis, the construction of new single-family homes was 38.5% less than the 1963-to-2008 average in 2015. Housing permits were estimated at a seasonally adjusted annualized rate (SAAR) of 1.17 million in April 2016, down by 1.9% from April 2015 (graph 11.4.2) (US Census Bureau, 2016a).

Sales of existing (i.e. previously owned) homes improved in 2015: 5.25 million existing homes were sold, compared with 4.94 million in 2014 – a 6.3% increase. The median existing-home sale price in April 2016 was $232,500, 6.3% higher than in April 2015 ($218,700) (National Association of Realtors, 2016). There were 591,000 new-home sales (i.e. sales of newly constructed homes) in 2015 (SAAR) (US Census Bureau, 2016c). The volume of new-home sales was similar to the volume in 1966-1970, when the civil noninstitutional population was about 132 million, compared with nearly 253 million in 2016 (Federal Reserve Bank of St. Louis, 2016); the number of built dwellings being built is insufficient for current population growth. The median new-home sale price was $321,100 in April 2016, up by 9.7% compared with April 2015 ($292,700). US average house size and price also increased, to 2,665 square feet and $379,800, respectively, in the first quarter of 2016 (US Census Bureau, 2016c,d). The increase in house prices – both new and existing – is raising concerns that housing affordability may be a problem in the future (Joint Center for Housing, 2016a).

GRAPH 11.4.2

US housing permits, starts and completions, 2011-2016

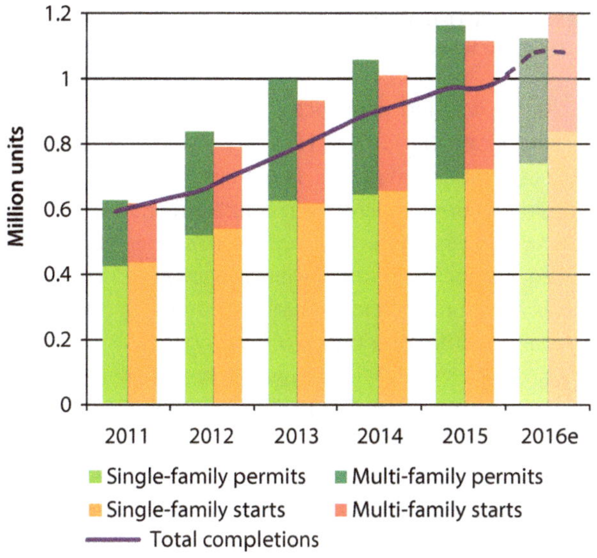

Notes: e = estimate (January-June 2016 data); SAAR = seasonally annualized adjusted rate.
Source: US Census Bureau, 2016a; Mortgage Bankers Association, 2016.

Total private residential construction spending (i.e. single-family + multifamily + remodelling) increased by 13.0% in 2015, year-on-year, to $418.3 billion (graph 11.4.3). Single-family construction spending increased by 12.9%, to $218.5 billion; multifamily expenditure increased by 24.1%, to $51.9 billion and house renovation spending increased by 9.2%, to $147.0 billion (all SAAR and nominal US dollars). Private non-residential spending increased by 4.0% in 2015, year-on-year, to $404.5 billion, and public expenditure increased by 2.2%, to $297.8 billion (US Census Bureau, 2016e). The Joint Center for Housing (2016b) projected that $309.8 billion (€277.0 billion) might be spent on remodelling in 2016.

GRAPH 11.4.3

US construction spending, 2006-2016

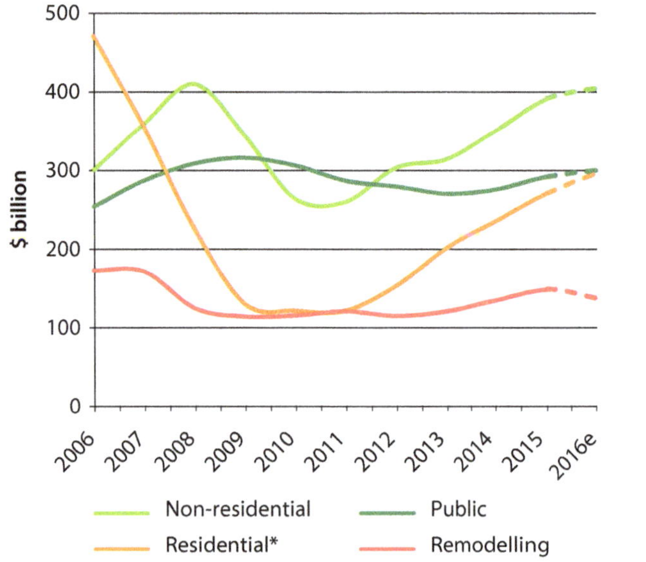

Notes: *Private residential spending less remodelling expenditure (SAAR); e = estimate (January-April 2016 data), and nominal values.
Source: US Census Bureau, 2016c.

Historically, US housing construction and sales have been a major component of US GDP. Housing includes: residential investment (construction of new single-family and multifamily homes, residential remodelling, the production of manufactured homes, and brokers' fees); housing services spending (rent, owner's equivalent rent, and utilities); and expenditure on furnishings and durable goods. Before the housing crash and the financial crisis, the contribution of housing to GDP averaged 17-19%; it was 15.3% in 2015, compared with 18.6% in 2005. Residential investment is the crucial component: it peaked in 2005 at 6.1% of GDP but was only 3.0% in 2014 and 3.2% in 2015, an indication that the new-home construction market has further room to expand (National Association of Homebuilders, 2016).

11.4.2 United States construction outlook

The US housing market has stabilized and all sectors have improved since 2009, the market's nadir. A more robust housing construction and sales market may be hindered, however, by a tendency for "millennials" (adults born in 1982 or later) to live with their parents; a relatively low level of household formation; student debt; underemployment and stagnant-to-declining median incomes; banking regulations (including stringent down-payment requirements and credit reporting); a constrained housing inventory; and a fragile economy. Changing attitudes towards home ownership have been reported, including the trend of renting rather than purchasing houses. Also mentioned are building regulations and the unavailability of building lots and construction workers. Numerous commentaries exist on the lack of starter houses being built for first-time buyers (Buehlmann and Alderman, 2016). Negative-equity or "underwater" homes are now at the same level as before the housing crash (CoreLogic, 2016).[27] In the past, houses with negative equity were viewed as hindrances to future building construction and the overall housing market.

The Mortgage Bankers Association (2016) projected housing starts in the US at 836,000 in 2016, 954,000 in 2017, and 1,063,000 in 2018. It also forecast sales of new single-family homes of 574,000 units in 2016, 669,000 in 2017, and 702,000 in 2018. Sales of existing homes are projected at 5,472,000 units in 2016, 5,768,000 in 2017, and 5,885,000 in 2018.

Source: APA, 2016.

11.4.3 Canadian housing construction market

The OECD (2016) projected Canadian economic growth to increase in 2016 and to achieve 2.2% in 2017. Marple and DePratto (2016) projected that real GDP growth would average 1.9% in 2016, 2.0% in 2017, 1.9% in 2018 and 1.7% in 2019. Concerns are being raised about escalating house prices in the Toronto and Vancouver markets; specifically, the concern is that the increase is unsustainable (Royal Bank of Canada, 2016; Blatchford, 2016).

The Canadian housing market is declining slightly: the estimate of 186,800 starts in 2016 is down from the 195,535 starts in 2015, and only 177,800 starts are expected in 2017. Of the projected starts in 2016, 71,800 are forecast as single-family units and 115,050 as multifamily; in 2017, the projection is for 65,200 single-detached and 114,050 multifamily unit starts (graph 11.4.4). Sales

[27] A negative-equity (or underwater) house is one in which the borrower owes more on the mortgages than the value of the house.

are forecast at between 501,700 and 525,400 units in 2016 and between 485,500 and 508,400 units in 2017 (CMHC, 2016).

Unemployment in Canada is forecast to increase to 7.2% in 2017 (CHMC, 2016), up from 6.8% (CHMC, 2015). Wages are projected to increase by 2.5% in 2016, which may aid the housing market (CHMC, 2016).

GRAPH 11.4.4

Housing starts, Canada, 2007-2017

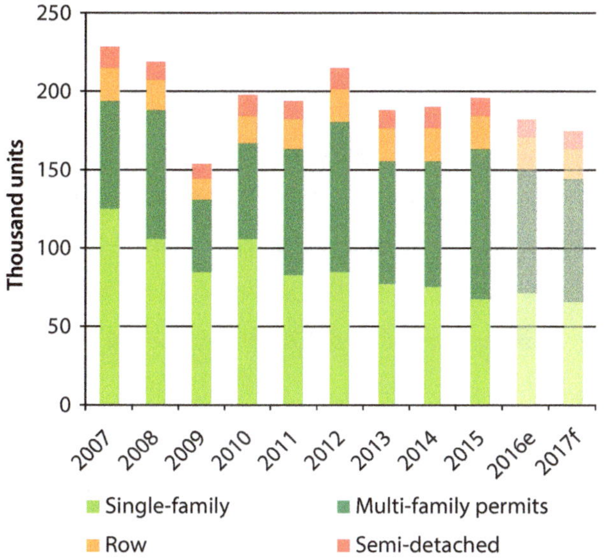

Notes: e = estimate; f = forecast.
Source: CMHC, 2016.

11.5 REFERENCES

Allen, T. 2016. Production in construction down by 0.9% in euro area. Available at: http://ec.europa.eu/eurostat/documents/2995521/7311581/4-19052016-AP-EN.pdf/ca7208ab-c306-4965-8104-e70d0a0f16ce

Blatchford, A. 2016. Bank of Canada issues stern warning on Toronto, Vancouver housing markets. The Canadian Press. Available at: www.ctvnews.ca/business/bank-of-canada-issues-stern-warning-on-toronto-vancouver-housing-markets-1.2938420

Buehlmann, U. and Alderman, D. 2016. The Virginia Tech – US Forest Service April 2016 housing commentary: sections I and II. Available at: http://woodproducts.sbio.vt.edu/housing-report/casa-2016-04a-april-main.pdf

CMHC. 2015. Housing market outlook. Canada Mortgage and Housing Corporation (CMHC). Available at: www.cmhc-schl.gc.ca/odpub/esub/61500/61500_2015_Q02.pdf

CMHC. 2016. Housing market outlook. Canada Mortgage and Housing Corporation (CMHC). Available at: www.cmhc-schl.gc.ca/odpub/esub/61500/61500_2016_Q02.pdf

CoreLogic. 2016. Corelogic reports 268,000 US homeowners regained equity in the first quarter of 2016. Available at: www.corelogic.com/about-us/news/corelogic-reports-268,000-us-homeowners-regained-equity-in-the-first-quarter-of-2016.aspx

Euroconstruct. 2009. 68th Euroconstruct summary report. 68th Euroconstruct Conference. November 2009. Zurich, Switzerland.

Euroconstruct. 2013. 75th Euroconstruct summary report. 75th Euroconstruct Conference. June 2013. Copenhagen, Denmark.

Euroconstruct. 2015. 80th Euroconstruct summary report. European construction: market trends until 2018. 80th Euroconstruct Conference. December 2015. Budapest, Hungary.

Eurostat. 2016. Housing statistics. Available at: http://ec.europa.eu/eurostat/statistics-explained/index.php/Housing_statistics

Federal Reserve Bank of St. Louis. 2016. Civilian noninstitutional population. Available at: https://fred.stlouisfed.org/series/CNP16OV

Federal State Statistics Service. 2016a. Current Statistical Survey Magazine, 1(96). Moscow. Available at: www.gks.ru/free_doc/doc_2016/magazine/st_ob1-eng.pdf

Federal State Statistics Service. 2016b. About housing 2015. Moscow. Available at: www.gks.ru/bgd/free/B04_03/IssWWW.exe/Stg/d06/19.htm

IMF. 2016. World economic outlook: too slow for too long. International Monetary Fund (IMF). Available at: www.imf.org/external/pubs/ft/weo/2016/01/

Joint Center for Housing. 2016a. The state of the nation's housing 2016. Available at: www.jchs.harvard.edu/sites/jchs.harvard.edu/files/jchs_2016_state_of_the_nations_housing_lowres.pdf

Joint Center for Housing. 2016b. Home remodeling activity looking to gain steam through mid-year. Available at: www.jchs.harvard.edu/home-remodeling-activity-looking-gain-steam-through-mid-year

Marple, J. and DePratto, B. 2016. Long-term economic forecast. 23 March. TD Economics. Available at: www.td.com/document/PDF/economics/qef/long_term_mar2016.pdf

Mortgage Bankers Association. 2016. MBA mortgage finance forecast. 14 July. Available at: www.mba.org/Documents/Research/Mtg%20Fin%20Forecast%20Jun%202016.pdf

National Association of Homebuilders. 2016. Housing's contribution to gross domestic product (GDP). Available at: www.nahb.org/en/research/housing-economics/housings-economic-impact/housings-contribution-to-gross-domestic-product-gdp.aspx

National Association of Realtors. 2016. April existing-home sales ease. Available at: www.realtor.org/news-releases/2011/05/april-existing-home-sales-ease

OECD. 2016. Canada. OECD economic outlook. Volume 2016(1). Preliminary version. Organisation for Economic Co-operation and Development (OECD). Available at: www.oecd.org/eco/outlook/economic-forecast-Canada-June-2016.pdf

PMR. 2016. Construction sector in Russia H1 2016. Market analysis and development forecasts for 2016-2021. Available at: www.cee-construction.com/analysis/1221/infographics-construction-sector-in-russia-h1-2016

Royal Bank of Canada. 2016. Canadian housing health check. May. RBC Economics. Available at: www.rbc.com/economics/economic-reports/pdf/canadian-housing/healthcheck-may16.pdf

US Census Bureau. 2016a. New residential construction in June 2016. Press release. Available at: https://www.census.gov/construction/nrc/pdf/newresconst.pdf

US Census Bureau. 2016b. New residential sales in April 2016. Available at: www.census.gov/construction/nrs/pdf/newressales_201604.pdf

US Census Bureau. 2016c. Median and average sale price of new homes sold in the United States. Available at: www.census.gov/construction/nrs/pdf/uspricemon.pdf

US Census Bureau. 2016d. New privately owned housing units started in the United States by purpose and design. Available at: www.census.gov/construction/nrc/pdf/quarterly_starts_completions.pdf

US Census Bureau. 2016e. Value of private construction put in place – seasonally adjusted annual rate. Available at: www.census.gov/construction/c30/pdf/privsa.pdf

World Bank. 2016. Global outlook: divergences and risks, global economic prospects. Available at: pubdocs.worldbank.org/pubdocs/publicdoc/2016/5/201061463685562570/Global-Economic-Prospects-June-2016-Highlights-Chapter-1.pdf

ANNEXES

COMPONENTS OF WOOD PRODUCTS GROUPS

(Based on Joint Forest Sector Questionnaire nomenclature)

The important breakdowns of the major groups of primary forest products are diagrammed below. In addition, many sub-items are further divided into softwood or hardwood. These are: all the roundwood products; sawnwood; veneer sheets; and plywood. Items that do not fit into listed aggregates are not shown. These are wood charcoal; wood chips and particles; wood residues; sawnwood; other pulp; and recovered paper. The sources for pictures used in these diagrams are databanks of Metsä Group (2012), Raunion Saha (2012), Stora Enso (2012) and UPM (2012).

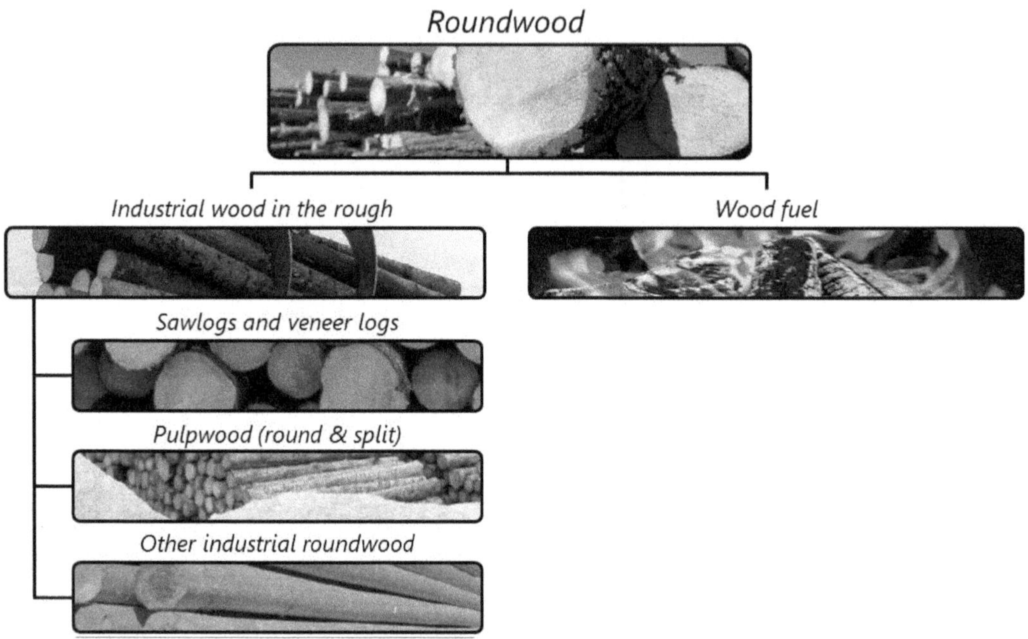

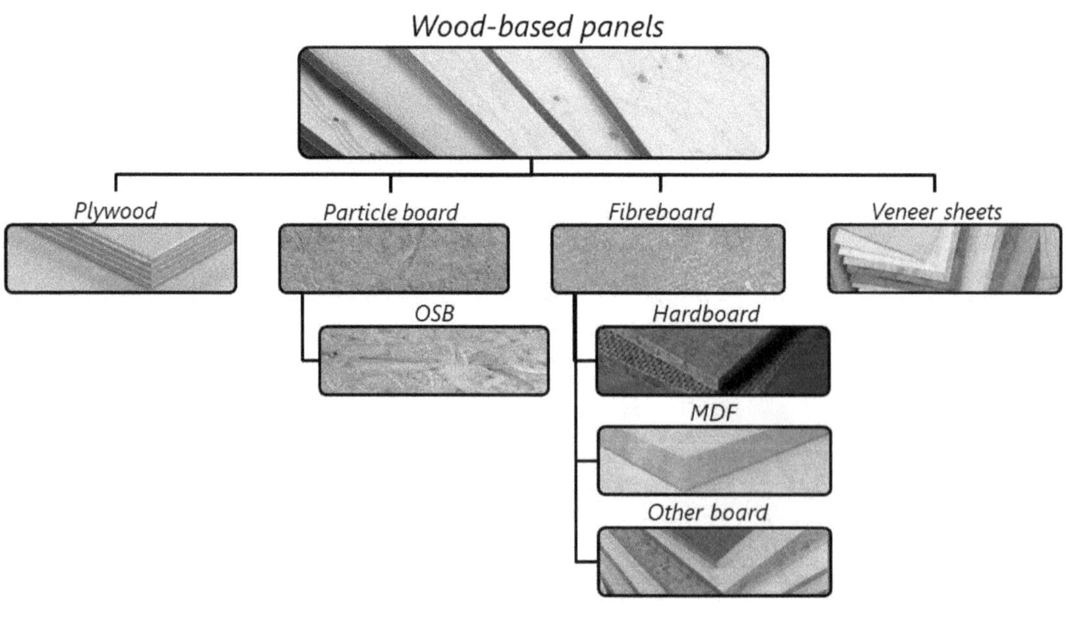

Woodpulp

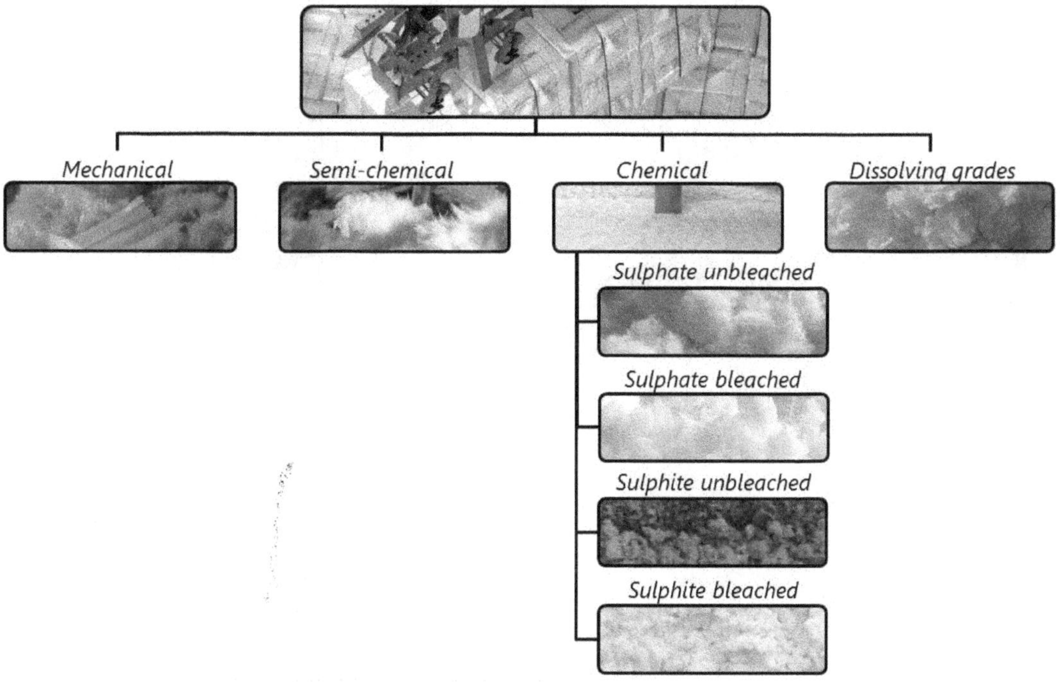

Paper and paperboard

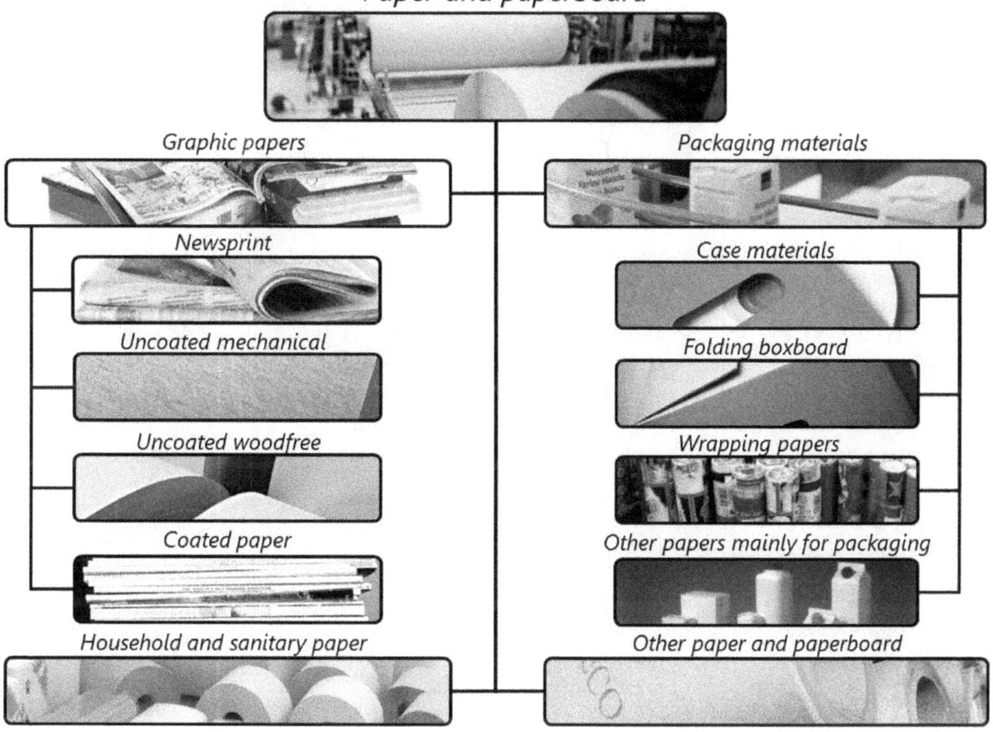

COUNTRIES IN THE UNECE REGION AND ITS SUBREGIONS

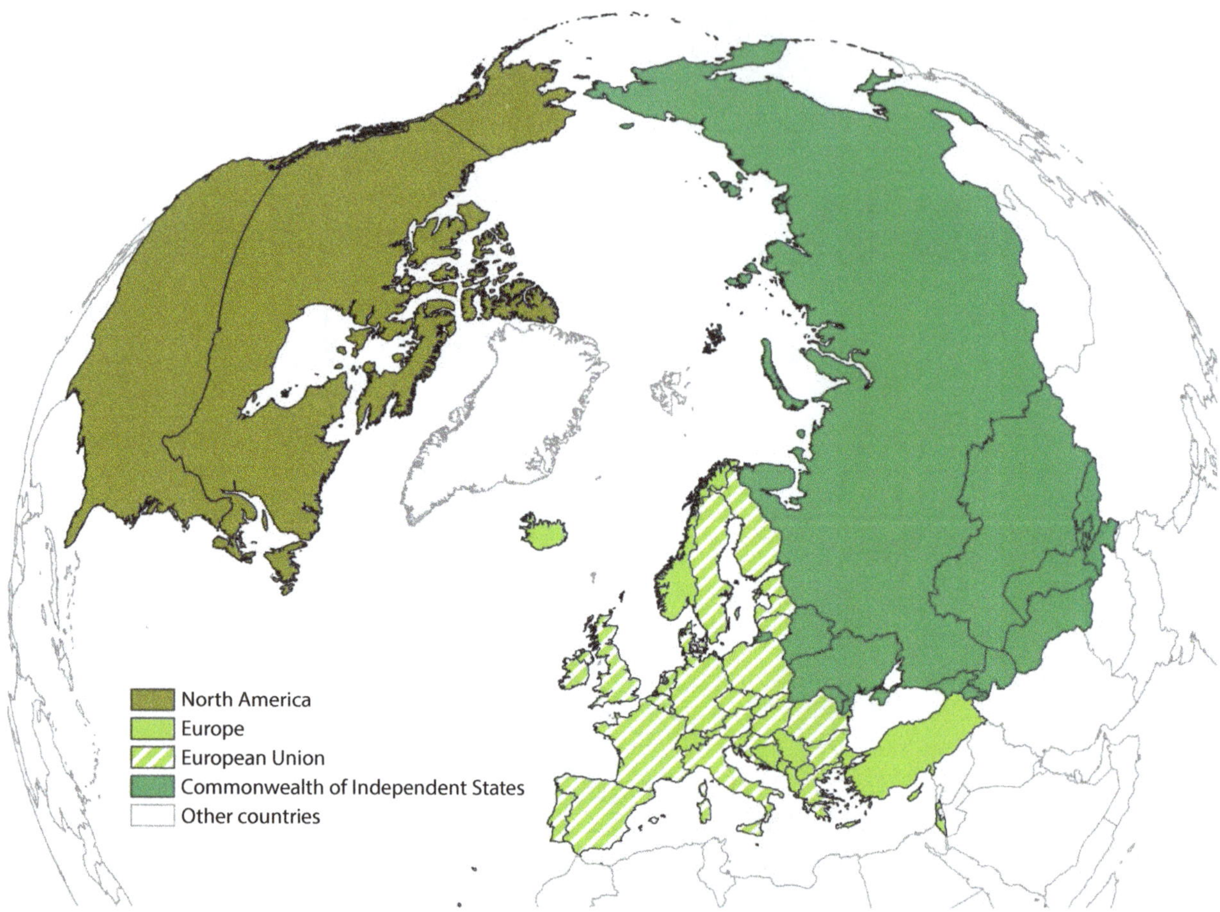

Commonwealth of Independent States
Armenia
Azerbaijan
Belarus
Georgia
Kazakhstan
Kyrgyzstan
Republic of Moldova
Russian Federation
Tajikistan
Turkmenistan
Ukraine
Uzbekistan

North America
Canada
United States of America

Europe		
European Union		*Other countries*
Austria	Italy	Albania
Belgium	Latvia	Andorra
Bulgaria	Lithuania	Bosnia and Herzegovina
Croatia	Luxembourg	Iceland
Cyprus	Malta	Israel
Czech Republic	Netherlands	Liechtenstein
Denmark	Poland	Monaco
Estonia	Portugal	Montenegro
Finland	Romania	Norway
France	Slovakia	San Marino
Germany	Slovenia	Serbia
Greece	Spain	Switzerland
Hungary	Sweden	The former Yugoslav Republic of Macedonia
Ireland	United Kingdom	Turkey

LIST OF AUTHORS

First Name	Family Name	Chapter	Affiliation	Email address
Karen	Abt	9	Southern Research Station, USDA Forest Service, Research Triangle Park, NC	kabt@fs.fed.us
Francisco	Aguilar	9	University of Missouri, Columbia, Missouri, US	aguilarf@missouri.edu
Eduard	Akim	8	State Technological University of Plant Polymers, St Petersburg, Russian Federation	akim-ed@mail.ru
Delton	Alderman	11	USDA Forest Service, Princeton, West Virginia, US	dalderman@fs.fed.us
Gunilla	Beyer	3	Swedish Forest Industries Federation	gunilla.beyer@skogsindustrierna.org
Ivan	Eastin	7	University of Washington, Seattle, Washington, US	eastin@uw.edu
Håkan	Ekström	4	Wood Resources International LLC, Seattle, Washington, US	hakan@woodprices.com
Kathryn	Fernholz	2	Dovetail Partners, Minneapolis, Minnesota, US	katie@dovetailinc.org
Christopher	Gaston	10	FPInnovations, Vancouver, Canada	chris.gaston@fpinnovations.ca
Branko	Glavonjić	9	Faculty of Forestry, Belgrade State University, Belgrade, Serbia	branko.glavonjic@sfb.bg.ac.rs
Antti	Koskinen	5	Pöyry Management Consulting Oy, Vantaa, Finland	antti.koskinen@poyry.fi
Florian	Kraxner	2	International Institute for Applied Systems Analysis, Laxenburg, Austria	kraxner@iiasa.ac.at
Marion	Le Roy	7	European Panel Federation, Brussels, Belgium	marion.leroy@europanels.org
Bernard	Lombard	8	CEPI-Confederation of European Paper Industries, Brussels, Belgium	b.lombard@cepi.org
Eugene	Lopatin	9	The Finnish Forest Research Institute, Joensuu, Finland	eugene.lopatin@luke.fi
Warren	Mabee	9	Queens University, Kingston, Ontario, Canada	warren.mabee@queensu.ca
Frances	Maplesden	5, 6, 7	Maplesden Consulting, Rotorua, New Zealand	fran_map@clear.net.nz
Igor	Novoselov	2, 5, 7	WhatWood, Moscow, Russian Federation	igor.novoselov@whatwood.ru
Rupert	Oliver	2, 6	Forest Industries Intelligence Ltd, Settle, UK	rjwoliver@btopenworld.com
Tapani	Pahkasalo	10	Quantum Global Investment Management Ltd, Zug, Switzerland	Tapani.Pahkasalo@quantumglobal.ch
José	Palacín	1	United Nations Economic Commission for Europe, Geneva, Switzerland	jose.palacin@unece.org
Russ	Taylor	5	International Wood Markets Group Inc., Vancouver, BC, Canada	retaylor@woodmarkets.com
Michel	Valois	8	Valois Vision Marketing, Vancouver, BC, Canada	michel@valoisvision.com
Marshall	White	3	Virginia Tech, Department of Sustainable Biomaterials, Blacksburg, VA, US	mswhite@vt.edu

SOME FACTS ABOUT THE COMMITTEE ON FORESTS AND THE FOREST INDUSTRY

The UNECE Committee on Forests and the Forest Industries is a principal subsidiary body of the UNECE (United Nations Economic Commission for Europe) based in Geneva. It constitutes a forum for cooperation and consultation between member countries on forestry, the forest industry and forest product matters. All countries of Europe, the Commonwealth of Independent States, the United States of America, Canada and Israel are members of the UNECE and participate in its work.

The UNECE Committee on Forests and the Forest Industries shall, within the context of sustainable development, provide member countries with the information and services needed for policymaking and decision-making with regard to their forest and forest industry sectors, including the trade and use of forest products and, where appropriate, will formulate recommendations addressed to member governments and interested organizations. To this end, it shall:

1. With the active participation of member countries, undertake short-, medium- and long-term analyses of developments in, and having an impact on, the sector, including those developments offering possibilities for the facilitation of international trade and for enhancing the protection of the environment;
2. In support of these analyses, collect, store and disseminate statistics relating to the sector, and carry out activities to improve their quality and comparability;
3. Provide the framework for cooperation e.g. by organising seminars, workshops and ad hoc meetings and setting up time-limited ad hoc groups, for the exchange of economic, environmental and technical information between governments and other institutions of member countries required for the development and implementation of policies leading to the sustainable development of the sector and to the protection of the environment in their respective countries;
4. Carry out tasks identified by the UNECE or the Committee on Forests and the Forest Industries as being of priority, including the facilitation of subregional cooperation and activities in support of the economies in transition of central and eastern Europe and of the countries of the region that are developing from an economic perspective;
5. It should also keep under review its structure and priorities and cooperate with other international and intergovernmental organizations active in the sector, and in particular with the FAO (the Food and Agriculture Organization of the United Nations) and its European Forestry Commission, and with the ILO (the International Labour Organisation), in order to ensure complementarity and to avoid duplication, thereby optimizing the use of resources.

More information about the Committee's work may be obtained by contacting:

UNECE/FAO Forestry and Timber Section
Forests, Land and Housing Division
United Nations Economic Commission for Europe/
Food and Agriculture Organization of the United Nations
Palais des Nations
CH-1211 Geneva 10, Switzerland

Fax: +41 22 917 0041
info.ECE-FAOforests@unece.org
www.unece.org/forests

UNECE/FAO PUBLICATIONS

Forest Products Annual Market Review 2015-2016 **ECE/TIM/SP/40**

Note: other market-related publications and information are available in electronic format at our website.

Geneva Timber and Forest Study Papers

Forest Products Annual Market Review 2014-2015	ECE/TIM/SP/39
Promoting Sustainable Building Materials and the Implications on the Use of Wood in Buildings	ECE/TIM/SP/38
Forests in the ECE Region: Trends and Challenges in Achieving the Global Objectives on Forests	ECE/TIM/SP/37
Forest Products Annual Market Review 2013-2014	ECE/TIM/SP/36
Rovaniemi Action Plan for the Forest Sector in a Green Economy	ECE/TIM/SP/35
The Value of Forests: Payments for Ecosystem Services in a Green Economy	ECE/TIM/SP/34
Forest Products Annual Market Review 2012-2013	ECE/TIM/SP/33
The Lviv Forum on Forests in a Green Economy	ECE/TIM/SP/32
Forests and Economic Development: A Driver for the Green Economy in the ECE Region	ECE/TIM/SP/31
Forest Products Annual Market Review 2011-2012	ECE/TIM/SP/30
The North American Forest Sector Outlook Study 2006-2030	ECE/TIM/SP/29
European Forest Sector Outlook Study 2010-2030	ECE/TIM/SP/28
Forest Products Annual Market Review 2010-2011	ECE/TIM/SP/27
Private Forest Ownership in Europe	ECE/TIM/SP/26
Forest Products Annual Market Review 2009-2010	ECE/TIM/SP/25
Forest Products Annual Market Review 2008-2009	ECE/TIM/SP/24
Forest Products Annual Market Review 2007-2008	ECE/TIM/SP/23
Forest Products Annual Market Review 2006-2007	ECE/TIM/SP/22
Forest Products Annual Market Review, 2005-2006	ECE/TIM/SP/21
European Forest Sector Outlook Study: 1960 – 2000 – 2020, Main Report	ECE/TIM/SP/20
Forest Policies and Institutions of Europe, 1998-2000	ECE/TIM/SP/19
Forest and Forest Products Country Profile: Russian Federation (Country profiles also exist on Albania, Armenia, Belarus, Bulgaria, former Czech and Slovak Federal Republic, Estonia, Georgia, Hungary, Lithuania, Poland, Romania, Republic of Moldova, Slovenia and Ukraine)	ECE/TIM/SP/18
Forest Resources of Europe, CIS, North America, Australia, Japan and New Zealand	ECE/TIM/SP/17

The above series of sales publications and subscriptions are available through United Nations Publications Offices as follows:

Sales and Marketing Section, Room DC2-853
United Nations
2 United Nations Plaza
New York, NY 10017
United States of America
Fax: + 1 212 963 3489
E-mail: publications@un.org
Web site: https://unp.un.org/

Geneva Timber and Forest Discussion Papers (original language only)

Forecast of the Committee on Forests and the Forest Industry: Forest Products Production and Trade 2014-2016	ECE/TIM/DP/64
Forecast of the Committee on Forests and the Forest Industry: Forest Products Production and Trade 2013-2015	ECE/TIM/DP/63
Competitiveness of the European Forest Sector	ECE/TIM/DP/62
Forecast of the Committee on Forests and the Forest Industry: Forest Products Production and Trade 2012-2014	ECE/TIM/DP/61
Forecast of the Committee on Forests and the Forest Industry: Forest Products Production and Trade 2011-2013	ECE/TIM/DP/60
Econometric Modelling and Projections of Wood Products Demand, Supply and Trade in Europe	ECE/TIM/DP/59
Swedish Forest Sector Outlook Study	ECE/TIM/DP/58
The Importance of China's Forest Products Markets to the UNECE Region	ECE/TIM/DP/57
Good Practice Guidance on Sustainable Mobilisation of Wood: Proceedings from the Grenoble Workshop	*ECE/TIM/DP/56
Harvested Wood Products in the Context of Climate Change Policies: Workshop Proceedings - 2008	*ECE/TIM/DP/55
The Forest Sector in the Green Economy	ECE/TIM/DP/54
National Wood Resources Balances: Workshop Proceedings	*ECE/TIM/DP/53
Potential Wood Supply in Europe	*ECE/TIM/DP/52
Wood Availability and Demand in Europe	*ECE/TIM/DP/51
Forest Products Conversion Factors for the UNECE Region	ECE/TIM/DP/49
Mobilizing Wood Resources: Can Europe's Forests Satisfy the Increasing Demand for Raw Material and Energy Under Sustainable Forest Management? Workshop Proceedings - January 2007	*ECE/TIM/DP/48
European Forest Sector Outlook Study: Trends 2000-2005 Compared to the EFSOS Scenarios	ECE/TIM/DP/47
Forest and Forest Products Country Profile; Tajikistan	*ECE/TIM/DP/46
Forest and Forest Products Country Profile: Uzbekistan	ECE/TIM/DP/45
Forest Certification – Do Governments Have a Role?	ECE/TIM/DP/44
International Forest Sector Institutions and Policy Instruments for Europe: A Source Book	ECE/TIM/DP/43
Forests, Wood and Energy: Policy Interactions	ECE/TIM/DP/42
Outlook for the Development of European Forest Resources	ECE/TIM/DP/41
Forest and Forest Products Country Profile: Serbia and Montenegro	ECE/TIM/DP/40
Forest Certification Update for the UNECE Region, 2003	ECE/TIM/DP/39
Forest and Forest Products Country Profile: Republic of Bulgaria	ECE/TIM/DP/38
Forest Legislation in Europe: How 23 Countries Approach the Obligation to Reforest, Public Access and Use of Non-Wood Forest Products	ECE/TIM/DP/37
Value-Added Wood Products Markets, 2001-2003	ECE/TIM/DP/36
Trends in the Tropical Timber Trade, 2002-2003	ECE/TIM/DP/35
Biological Diversity, Tree Species Composition and Environmental Protection in the Regional FRA-2000	ECE/TIM/DP/33
Forestry and Forest Products Country Profile: Ukraine	ECE/TIM/DP/32
The Development of European Forest Resources, 1950 To 2000: a Better Information Base	ECE/TIM/DP/31
Modelling and Projections of Forest Products Demand, Supply and Trade in Europe	ECE/TIM/DP/30
Employment Trends and Prospects in the European Forest Sector	ECE/TIM/DP/29
Forestry Cooperation with Countries in Transition	ECE/TIM/DP/28
Russian Federation Forest Sector Outlook Study	ECE/TIM/DP/27
Forest and Forest Products Country Profile: Georgia	ECE/TIM/DP/26

Forest certification update for the UNECE region, summer 2002	ECE/TIM/DP/25
Forecasts of economic growth in OECD and central and eastern European countries for the period 2000-2040	ECE/TIM/DP/24
Forest Certification update for the UNECE Region, summer 2001	ECE/TIM/DP/23
Structural, Compositional and Functional Aspects of Forest Biodiversity in Europe	ECE/TIM/DP/22
Markets for secondary processed wood products, 1990-2000	ECE/TIM/DP/21
Forest certification update for the UNECE Region, summer 2000	ECE/TIM/DP/20
Trade and environment issues in the forest and forest products sector	ECE/TIM/DP/19
Multiple use forestry	ECE/TIM/DP/18
Forest certification update for the UNECE Region, summer 1999	ECE/TIM/DP/17
A summary of "The competitive climate for wood products and paper packaging: the factors causing substitution with emphasis on environmental promotions"	ECE/TIM/DP/16
Recycling, energy and market interactions	ECE/TIM/DP/15
The status of forest certification in the UNECE region	ECE/TIM/DP/14
The role of women on forest properties in Haute-Savoie (France): Initial research	ECE/TIM/DP/13
Interim report on the Implementation of Resolution H3 of the Helsinki Ministerial Conference on the protection of forests in Europe (Results of the second enquiry)	ECE/TIM/DP/12
Manual on acute forest damage	ECE/TIM/DP/7

** signifies electronic publication only*

The above series of publications may be requested free of charge through:

UNECE/FAO Forestry and Timber Section
Forest, Land and Housing Division
United Nations Economic Commission for Europe
Palais des Nations
CH-1211 Geneva 10, Switzerland

Fax: +41 22 917 0041
E-mail: info.ECE-FAOforests@unece.org

Downloads are available at: www.unece.org/forests

www.ingramcontent.com/pod-product-compliance
Lightning Source LLC
Chambersburg PA
CBHW082347230926
51846CB00004B/940